Handbook of Manufacturing Systems Engineering

Handbook of Manufacturing Systems Engineering

Contributors

Ernesto López-Mellado et al.

AURIS
Reference

www.aurisreference.com

Handbook of Manufacturing Systems Engineering

Contributors: Ernesto López-Mellado et al.

Published by Auris Reference Limited

www.aurisreference.com

United Kingdom

Handbook of Manufacturing Systems Engineering

ISBN: 978-1-78154-927-8

British Library Cataloguing in Publication Data

A CIP record for this book is available from the British Library

Printed in the United Kingdom

Exclusively distributed by CBS Publishers & Distributors Pvt. Ltd.

Sales & Distribution Rights only for India, Pakistan, Bangladesh, Sri Lanka, Nepal and Bhutan.This book is not to be sold outside these territories.

Contents

List of Abbreviations

FPG	Flow of Parts Graph
MCS	Manufacturing Control System
KPCs	Key Product Characteristics
WIP	Work in Process
PVC	Polyvinyl Chloride
MS	Manufacturing Automation System
MDE	Model Driven Software Engineering
NAS	Networked Automation Systems
CNC	Computer Numerical Control
FMS	Flexible Manufacturing System
FRP	Fiber Reinforced Plastic
VARTM	Vacuum Assisted Rein Transfer Molding
EPSRC	Engineering and Physical Science Research Council
OMG	Object Management Group
CMP	Chemical Mechanical Polishing
EE	Equipment Engineer
FTIR	Fourier Transform Infrared Spectroscopy
GRnR	Gage Repeatability and Reproducibility
OPC	Optical Particle Counters
OCAP	Out of Control Action Plan
SEM	Scanning Electron Microscope
STEM	Systematic Machine Excursion Monitoring
VLSI	Very Large Scale Integration
CHP	Combined Heat and Power
PHR	Power-to-Heat Ratio
SHP	Separate Heating and Power
PGU	Power Generation Unit
SME	Shape Memory Effect
TTR	Transformation Temperatures
PWM	Pulse-Width Modulation

List of Contributors

Ernesto López-Mellado
CINVESTAV Unidad Guadalajara, Zapopan, Mexico

Yihai He
School of Reliability and Systems Engineering, Beihang University, Beijing 100191, China
Department of Systems Engineering and Engineering Management, City University of Hong Kong, Kowloon, Hong Kong

Zhenzhen He
School of Reliability and Systems Engineering, Beihang University, Beijing 100191, China

Linbo Wang
School of Reliability and Systems Engineering, Beihang University, Beijing 100191, China

Changchao Gu
School of Reliability and Systems Engineering, Beihang University, Beijing 100191, China

Kassu Jilcha
Addis Ababa University, Addis Ababa Institute of Technology, Addis Ababa, Ethiopia

Esheitie Berhan
Addis Ababa University, Addis Ababa Institute of Technology, Addis Ababa, Ethiopia

Hannan Sherif
Addis Ababa University, Addis Ababa Institute of Technology, Addis Ababa, Ethiopia

Birgit Vogel-Heuser
Institute of Automation and Information Systems, Technische Universität München, Munich, Germany

Ranbir Singh
Research Scholar, Department of Mechanical Engineering, DCRUST Murthal, Sonepat, India

Rajender Singh
Professor, Department of Mechanical Engineering, DCRUST Murthal, Sonepat, India

B. K. Khan
Technical Advisor, MSIT, Sonepat, India

Ya-Jung Lee
Engineering Science and Ocean Engineering, National Taiwan University, Chinese Taipei

Yu-Ti Jhan
Engineering Science and Ocean Engineering, National Taiwan University, Chinese Taipei

Cheng-Hsien Chung
R & D Department, United Ship Design & Development Center, Chinese Taipei

Yao Hsu
Business and Entrepreneurial Management, Kai Nan University, Taoyuan, Chinese Taipei

J M Edwards
MSI Research Institute, Department of Manufacturing Engineering, Loughborough University of Technology, Loughborough, Leicestershire, LE11 3TU, UK

I A Coutts
MSI Research Institute, Department of Manufacturing Engineering, Loughborough University of Technology, Loughborough, Leicestershire, LE11 3TU, UK

Faieza Abdul Aziz
Universiti Putra Malaysia Malaysia

Izham Hazizi Ahmad
Universiti Putra Malaysia Malaysia

Norzima Zulkifli
Universiti Putra Malaysia Malaysia

Rosnah Mohd. Yusuff
Universiti Putra Malaysia Malaysia

Chad A. Wheeley
Department of Mechanical Engineering, Mississippi St at e University, Oktibbeha County, USA

Pedro J. Mago
Department of Mechanical Engineering, Mississippi St at e University, Oktibbeha County, USA

Rogelio Luck
Department of Mechanical Engineering, Mississippi St at e University, Oktibbeha County, USA

Jorge Cortés
Tecnológico de Monterrey, Campus Monterrey México

Ignacio Varela-Jiménez
Tecnológico de Monterrey, Campus Monterrey México

Miguel Bueno-Vives
Tecnológico de Monterrey, Campus Monterrey México

Minna Lanz,
Tampere University of Technology Finland

Eeva Jarvenpaa,
Tampere University of Technology Finland

Fernando Garcia,
Tampere University of Technology Finland

Pasi Luostarinen
Tampere University of Technology Finland

Reijo Tuokko
Tampere University of Technology Finland

Preface

Handbook of Industrial and Systems Engineering is designed to provide students with the knowledge, skills, and abilities to successfully meet the most difficult challenges of modern manufacturing industries on a global scale. First chapter focuses on agent-based synthesis of distributed controllers for discrete manufacturing systems. Second chapter presents an approach to model the manufacturing system reliability dynamically based on their operation data of process quality and output data of product reliability. The objective of third chapter is to provide industry with a complete set of tools, techniques and procedures to allow examining of existing workers and machine performance systems. Fourth chapter presents factors, constraint and practical experience for the development of further usability experiments. Fifth chapter analyzes the research gaps, approach and techniques used, scope of new optimization techniques, objectives considered and validation approaches for loading problems of production planning in FMS. Sixth chapter focuses on deriving the in-plane permeability prediction method for Fiber Reinforced Plastics (FRP) laminates in the VARTM process by experimental measurements and numerical analysis. Seventh chapter describes a framework for building distributed manufacturing processes based on an integrating infrastructure. Through a decomposition based on application function, application interoperation and application interaction, proposals are made for structuring the integration software required to flexibly implement distributed systems which can evolve to support required change. In eighth chapter, a thorough investigation was carried out to improve shut down event problem at Metal Deposition process during wafer fabrication. Particle contamination on wafer surface can cause the circuit to malfunction and leading to machine shut down. The purpose of ninth chapter is to clearly outline a methodology to determine the economic effectiveness of installation and operation of a CHP system at industrial facilities that have a need for space or process heating in the form of steam. Tenth chapter focuses on reconfigurable tooling by using a reconfigurable material. Eleventh chapter discusses the possibilities of a modular and more transparent knowledge management concept that provides means for representing and capturing needed information as feasible as possible while understanding that it is also the software systems that need to adapt to the changes along the physical production systems.

Chapter 1

AGENT-BASED SYNTHESIS OF DISTRIBUTED CONTROLLERS FOR DISCRETE MANUFACTURING SYSTEMS

Ernesto López-Mellado

CINVESTAV Unidad Guadalajara, Zapopan, Mexico.

ABSTRACT

A method for designing real-time distributed controllers of discrete manufacturing systems is presented. The approach held is agent based; the controller strategy is distributed into several interacting agents that operate each one on a part of the manufacturing process; these agents may be distributed into several interconnected processors. The proposed method consists of a modeling methodology and software development framework that provides a generic agent architecture and communication facilities supporting the interaction among agents.

INTRODUCTION

Nowadays discrete manufacturing systems are large and complex systems that integrate several kinds of devices of miscellaneous nature and behavior, namely robots, conveyors, machines, sensors, etc. Additionally, the production requirements are often changed; these facts impose to the system components to be versatile, and to the coordination system or controller to be highly flexible. The core of a control system is complex software, which determines at last, the flexibility and the performance of the automated system.

This control software provides several functions: tasks execution, monitoring, decision making, and planning; it is generally distributed into a four layer hierarchy [1]; in this scheme the control function is decomposed into four levels in which the response time is shortest in the lower levels. The lowest level includes the local controllers of the physical devices in the cell (robots, conveyors, machines, sensors, etc. The task coordination level or cell level manages and supervises the activities of the local controllers involved in a cell by the generation of pertinent commands according to events issued from the local control level.

The task planner generates the strategy of the controller for the cells according to the specifications from the production planning level. Due to the complexity of the tasks, especially which performed at the cell level, the synthesis of the controller is a difficult job often addressed through planning techniques. One of the problems found in the synthesis realtime controllers is that the representation of the qualitative controller often includes a large amount of knowledge whose processing is time consuming. This work deals with the task coordinator level. The functions of this control level can be decomposed mainly in a) the sequencing of operations to accomplish the assembly task in a normal functioning regime, and b) the handling of exceptions representing operation failures [2]. In this paper the case of normal execution is addressed.

The design and implementation of tasks controllers of complex manufacturing systems has been addressed by using several approaches. The object oriented approach has been held for modelling [3], simulation [4-6], and control [7] of manufacturing systems. The agent based approach [8] has been adopted to address some problems in manufacturing systems [9]; DeLoach [10] proposes a modelling language for describing the diverse kinds of agents, and defines a methodology (MaSE) for the formal synthesis of agent systems; in [11] Bussmann focuses on decision making issues during the planning stage, and in [12] he addresses the task programming issue proposing a synthesis method that leads to concurrent centralised control software; in [13] Ouelhadj proposed a dynamic control architecture for manufacturing systems organised into cells, but the programming of the agents is not reported.

In this work we also profit of the agent-based approach for conceiving a control software as a composition of interacting modules, defined as reactive agents, which may be distributed into several processors; it is proposed a method that supports the complete development life cycle of distributed controllers of discrete manufacturing systems; this lifecycle is shown in Figure 1, in which the stages of the method are pictorially overviewed. The proposed method for sequencing the activities of the cell components allows building rapidly prototypes of software controllers.

The remainder of this paper is organized as follows: Section 2 describes the proposed methodology for modeling both the manufacturing system and the tasks to be executed; Section 3 presents the proposed method for the design of distributed control software: first the decomposition of the task model is described, then agent based solution to the synthesis of distributed software is outlined

Controller Modeling

This section presents the methodology that helps to obtain systematically the contents of the knowledge bases from a model of the manufacturing/assembly system; this model, close to that presented in [7], includes the system description and the tasks specification. The aim of this stage is to obtain in a structured way the system functioning and production requirements.

The description consists in a component classification of the manufacturing process, and a structuring of the system workspace.

The component classification leads to a taxonomy of the system components organised as a hierarchy including capabilities and features of each component, and the total quantity of devices. The hierarchy is useful to program the necessary classes in the control software. Figure 2 shows an example of components hierarchy.

System Description

The structuring of the system workspace consist of a definition of relevant physical emplacements where operations are performed on the work pieces or parts; rather than space partition into regions, the structuring is a discrete assignment of positions. The key element for structuring the workspace is the notion of site, which is defined as the place where parts can be temporary held or stored in a stable position (a table, a magazine, a robot gripper, ...) [2].

The sites may be single or composed (macro-site); single sites held one part or subassembly; composed sites have two or more emplacements which manage the information attached to a set of sites closely located and functionally equivalents. The sites that are associated to effectors are named active sites; otherwise they are called passive sites. A site contains information about the work zone were it is emplaced that can be used as mutual exclusion resource (for robot collision avoidance, for example).

Task Specification

The flow of material is described by a flow of parts graph (FPG), and then a set of dispatching rules, which represent the controller strategy, is obtained.

The Flow of Parts Graph

A FPG is a directed graph whose nodes are all the sites of the system; the arcs joining the nodes represent either the operations needed to transfer the parts from one site to another one or to modify the properties of the part held into a site. For example, Figure 3 describes the operations pick and place performed by a robot (R1); three sites are involved: two passive sites (CONV2 and TAB1), and an active site (GRIP1) ssociated to the robot gripper. The definition of FPG is given below:

Definition: A flow of parts graph is the tuple F = (G, SITES, OPER, e, l, f), where

- G is a connected directed graph G = (V, A), where —V is a finite set of vertex, —A Í V ′ V is a set of edges or arcs.
- SITES = {site1, ···, siten} is a finite set of site names, which are not input or output sites.
- e = {sitein1, ···, siteinp, siteout1, ···, siteoutq} is a finite set of site names labelling sites where the parts entry or leave the FMS.
- l: V ® SITES È {ε} is a labelling function that assigns name sites to the vertex of G.
- f: A ® OPER is a labelling function that assigns operations names to the arcs of G.

Sequencing the Operation

The dispatching rules are antecedent-consequent rules that state the conditions in which an operation must be executed; they are obtained directly from the FPG, and the number of rules is the same than the number of arcs in the FPG. The antecedent part is composed by conditions that involve mainly sensory conditions and tests (contents, part posture, ...) on the sites related by the operation; other conditions may involve tests on sites located upstream the FPG. The consequent part includes the request of execution of the associated operation and the updating of the involved sites.

A Modelling Example

For illustrating purposes we include an example regarding a simple assembly system; it will be addressed through the rest of the paper.

System and Task Description

1) *The system*: Consider the assembly cell sketched in Figure 4; it consist of three conveyor belts B1, B2, and B3, two robots R1, R2, two assembly tables A1, A2, an storing table ST. Each assembly table has two positions; the storing table has four positions. The parts to be assembled arrive through the conveyor B1; two kinds of assembled products the leave the cell through B2 and B3. In the front of R1, an optical sensor C1 detects

the arrival of parts; over this zone a camera of a location-recognition system is emplaced.

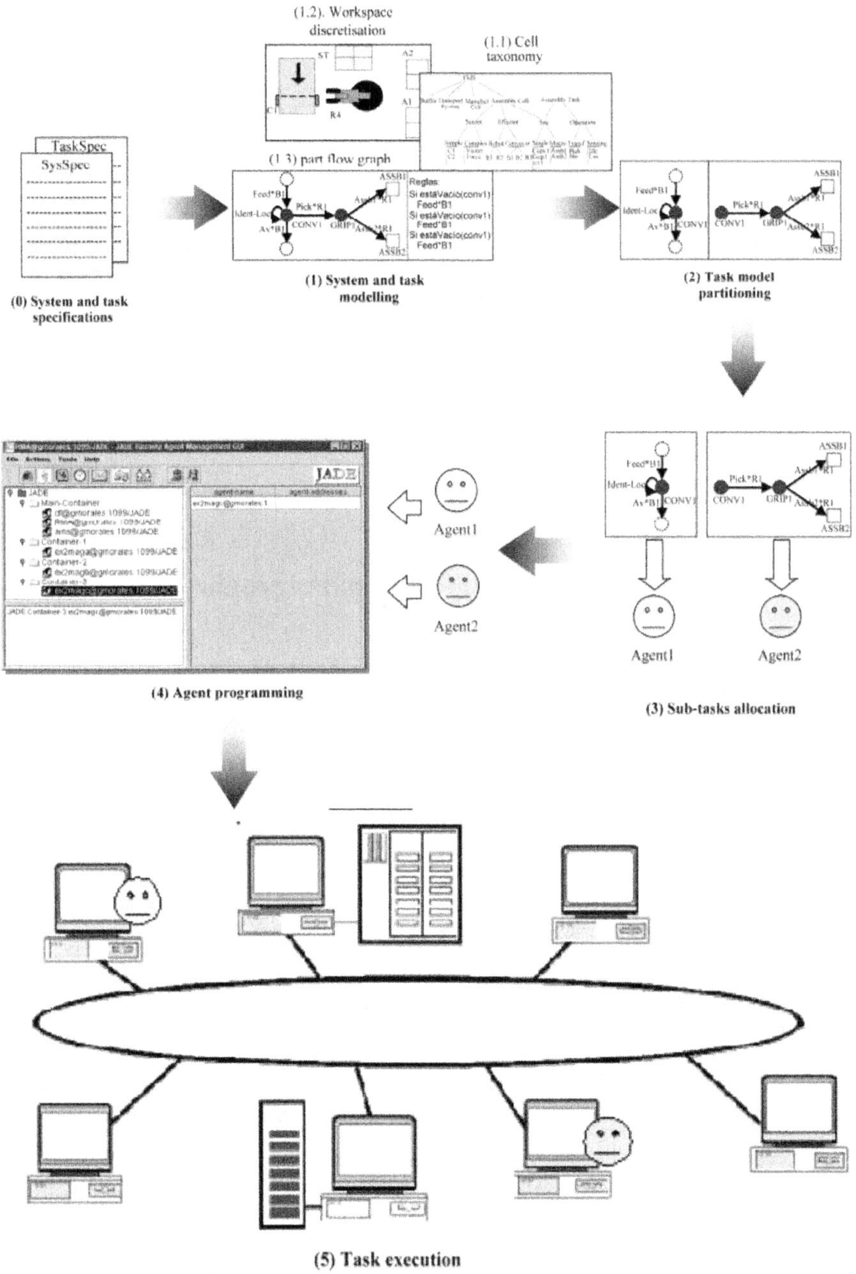

Figure 1: Software development life cycle.

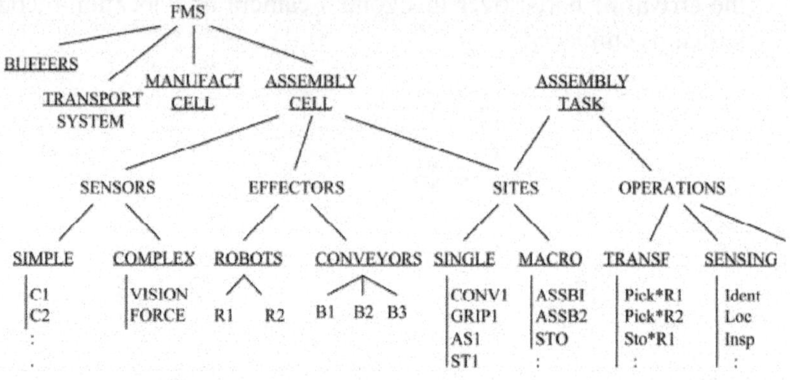

Figure 2: Taxonomy of the assembly cell.

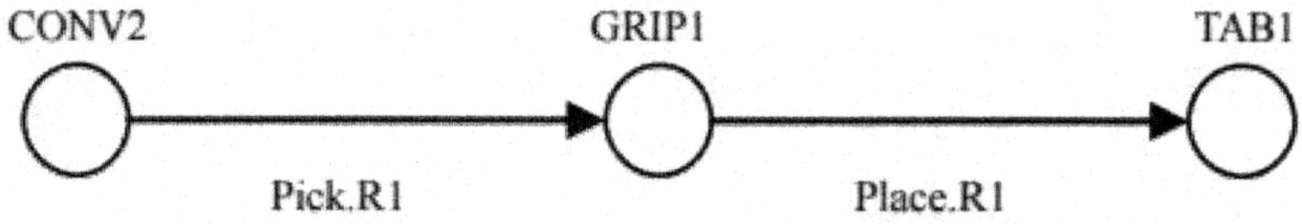

Figure 3. Flow of parts graph.

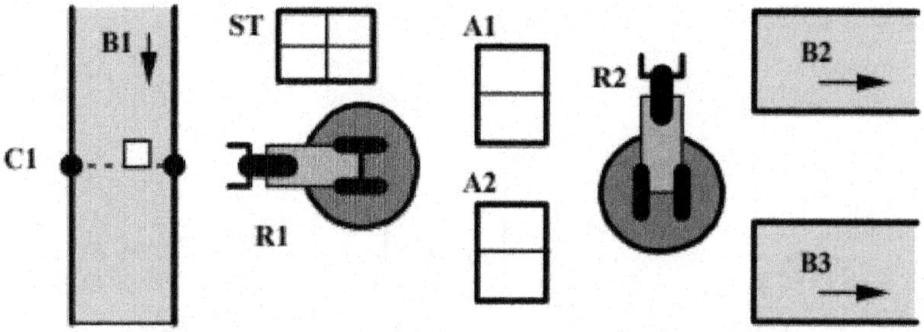

Figure 4: Assembly cell layout.

2) *The task:* Eight types of parts (A, B, C, D, E, F, G, and H) constitute the input flow in B1; they arrive at random order. R1 gets the parts and builds assemblies (stacks) on A1 (with the parts A, B, C, and D) or A2 (with the parts E, F, G, and H) according to a predefined order for each product. The detection of parts in C1 stops B1; the identity of the parts is determined by the vision system when they arrive at the position C1. R1 gets part only if it can be assembled or temporary stored in STi. Otherwise the part is left in B1. R2 gets completed assemblies from A1 or A2 and places them on B2 and B3 respectively.

System and Task Modelling

1) *System taxonomy.* The components of the assembly system are classed from a functional point of view: sensors, effectors, etc. This classification is useful to structure the factual knowledge of the assembly system: task state, component capabilities and relationships, etc. Figure 2 shows the hierarchy concerning the assembly system of the example; the items in the lowest level of the hierarchy can be object instances of the upper concept (class).

2) *Workspace modelling.* In the example the following sites are defined: CONV1, CONV2 and CONV3 are the sites associated to the place where the parts stops in front of the robot; GRIP1 and GRIP2 are associated to the grippers of R1 and R2 respectively. The storing table has four sites: STi (i = 1, ···, 4); they can be managed by the macro site ST. The sites associated to assembly tables are ASSB1 and ASSB2.

3) *Flow of parts.* The FPG shown in Figure 5 describes the flow of parts required in the assembly task; the sites defined in the model are related by the operations whose outcome is, mainly, the transferring the parts between the sites. Operations may only modify the properties of the part into a site; as an example notice that the operation Ident-Loc does not transfer the part to another site but it changes the attributes of the unknown part.

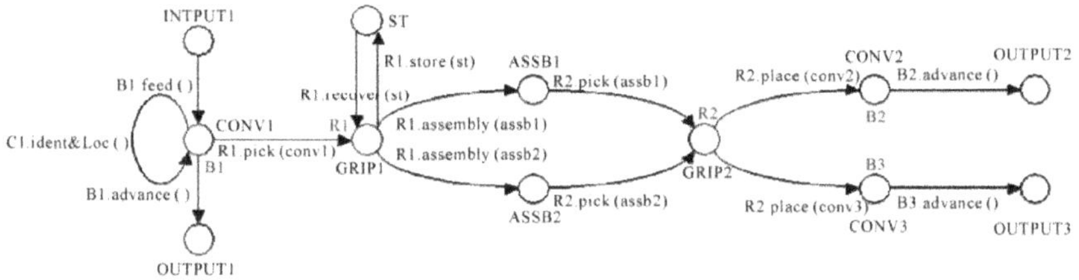

Figure 5: Flow of parts graphs for the assembly cell.

Operations may also put parts into the flow model or drawn out parts (or products) from the model. The FPG for this example is defined as follows

F = (G, SITES, OPER, λ, ϕ) where G = (V, A) with

V = {v_0, v_1, v_2, v_3, v_4, v_5, v_6, v_7, v_8, v_9, v_{10}, v_{11}}

A = {(v_0, v_1), (v_1, v_2), (v_1, v_1), (v_1, v_3), (v_3, v_4), (v_4, v_3), (v_3, v_5), (v_3, v_6), (v_5, v_7), (v_6, v_7), (v_7, v_8), (v_7, v_9), (v_8, v_{10}), (v_8, v_{11})}

SITES = {conv1, conv2, conv3, grip1, grip2, assb1, assb2, st1}

ε = {input1, output1, output2, output3}

OPER = {B1.feed(), B1.advance(), R1.pick(conv1), R1. store(st), R1.recover(st), R1.assembly(assb1), R1. assembly(assb2), R2.pick(assb1), R2.pick(assb2), R2.place-(conv2), R2.place2(conv3), B2.advance(), B3.advance(), C1.ident&Loc()}

λ = {(v_0, input1), (v_1, conv1), (v_2, output1), (v_3, grip1), (v_4, st1), (v_5, assb1), (v_6, assb2), (v_7, grip2), (v_8, conv2), (v_9, conv3), (v_{10}, output2), (v_{11}, output3)}

ϕ = { ((v_0, v_1), B1.feed()), ((v_1, v_2), B1.advance()), ((v_1, v_3), R1.pick(conv1)), ((v_3, v_4), R1.store(st)), ((v_4, v_3), R1.retreive(st)), ((v_3, v_5), R1.assembly(assb1)), ((v_3, v_6), R1.assembly(assb2)), ((v_5, v_7), R2.pick(assb1),), ((v_6, v_7), R2.pick(assb2)), ((v_7, v_8), R2.place(conv2)), ((v_7, v_9), R2.place(conv3)), ((v_8, v_{10}), B2.advance()), ((v_8, v_{11}), B3.advance()), ((v1,v1), C1.ident&Loc())}

Distributed Software Design

This section deals with the design of the distributed software that implements the task controller of a manufacturing system. First the modularisation of the task model is presented, and then the resulting partition is taken for implementing the agents [14].

Task Model Partition

The task model must be decomposed into subtasks in such manner that every subtask may be assigned to an agent; this decomposition is achieved by a partition of the FPG. Several strategies for obtaining sub-graphs from the FPG may be adopted according to the number of processors, the geographical distribution on the components, or the similarity of the sub-graphs.

In this work the strategy held for decomposing the graph is creating the maximal number of sub-graphs; each sub-graph must involve one active site. This approach allows defining agents capable to control one effector. A three-step algorithm is described below.

Algorithm: Partitioning the task model.

1) *Identify active and passive sites.* Let Act Í SITES, the set of active sites and Pass Í SITES the set of passive sites.
2) *Sub-model creation.* For every sÎAct, create a subgraph gk including s and its predecessors and successors. In gk it is included an active site and several passive sites. SG = {g1, g2, ···, gr} is the set of sub-models.
3) *Simplification of sub-models.* The operations associated to the arcs of a graph gk must be executed by the effectors or sensors associated to the active site of gk. Thus the arcs labelled with other operations must be withdrawn from gk; consequently isolated vertex must be eliminated too.

The sites belonging to two or more gk are called interface sites; they are in the boundary of the graph and they must be carefully managed because they are considered as shared resources.

Example: Consider the FPG of Figure 5; defining Act = {$conv_1$, $conv_2$, $conv_3$, $grip_1$, $grip_2$}, and Pas = {asb_1, $assb_2$, st}, the decomposition procedure yields the subgraphs depicted in Figure 6; so SG = {g_1, g_2, g_3, g_4, g_5}.

Task Programming Framework

Once the task model is decomposed, each sub-model is used for defining an agent. Since every sub-model involves an actuator, it must be controlled by the corresponding agent avoiding situations in which two or more agents handle the same effector. The knowledge base of every agent corresponds to the set of rules associated to the arcs of the pertaining subtask

The developed platform supports the programming of the agents that implement the subtasks of the system; this platform is based on JADE V1.2 (Java Agent Development framework) [15], which meets the standards of FIPA [16].

Requirements

Agent Requirements: Each agent

- has a unique name
- manages the corresponding subtask; it initializes and updates the contents of its sites and macro-sites.

- exchanges messages with other agents; it interprets the message and perform the requests such as provide information about the contents of a site or about the execution of an operation.
- coordinates with other agents for allocating shared resources
- manages the associated devices (effectors and sensors involved in the subtask).

Sites requirement:. Each site or macro-site

- has a unique name.
- provides facilities for managing its state, i.e. the initializing and updating of the associated attributes such as contents, position, sensor values, trace of the operations performed on it, etc.
- provides facilities for managing special features, such as flags for mutual exclusion, the agent names that share it, or the assembly patterns if it is a site for assembly.

Devices requirements: For every device (effector or sensor) in the system, one must create an interface module that allows consulting the device state and handling the messages representing actions requests or responses. Every module has a unique name.

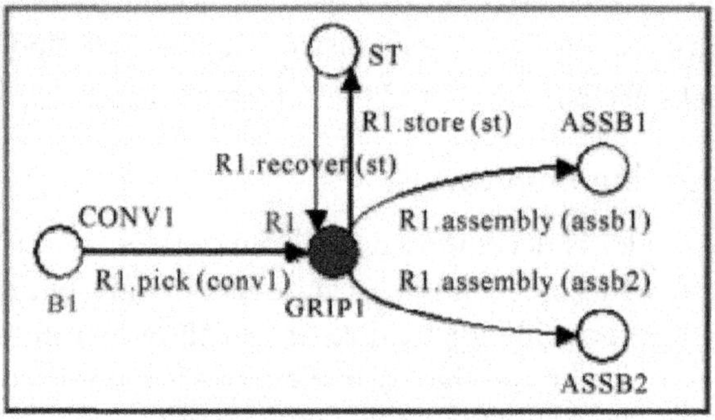

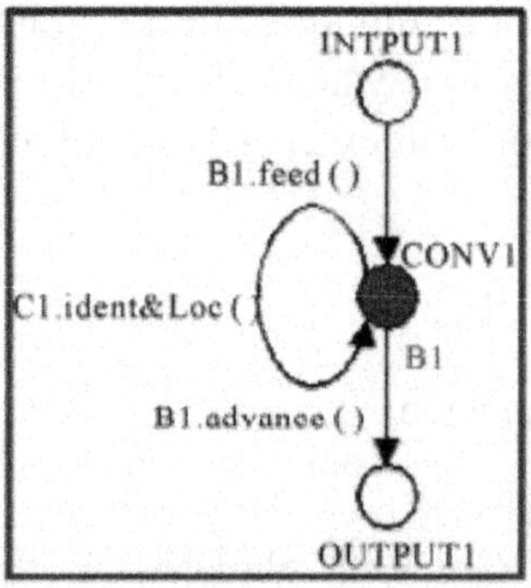

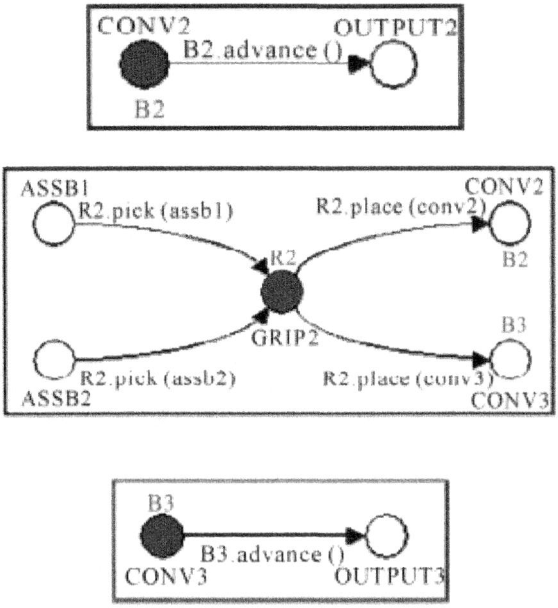

Figure 6: Graph decomposition.

Definition of Classes

Component architecture: The implemented components are integrated into a package organised in four sub-packages; it is shown in Figure 7. The sub-package MApackage sites contains the classes Site and Part, MApackage macrosites contains the class Macrosite and the classes related with the access of sites contained in the macrosite. Three kinds of sites are considered: interface sites (shared by two or more sub-tasks), internal sites, and remote sites (none shared but belonging to other sub-tasks). Figure 8 shows a detail of the above packages.

The sub-package MApackage devices contains the class Device and other sub-packages. MApackage devices effectors and MApackage devices sensors contain subclasses illustrated in Figure 9. This organization is strongly suggested by the taxonomy of the manufacturing system obtained during the modeling stage.

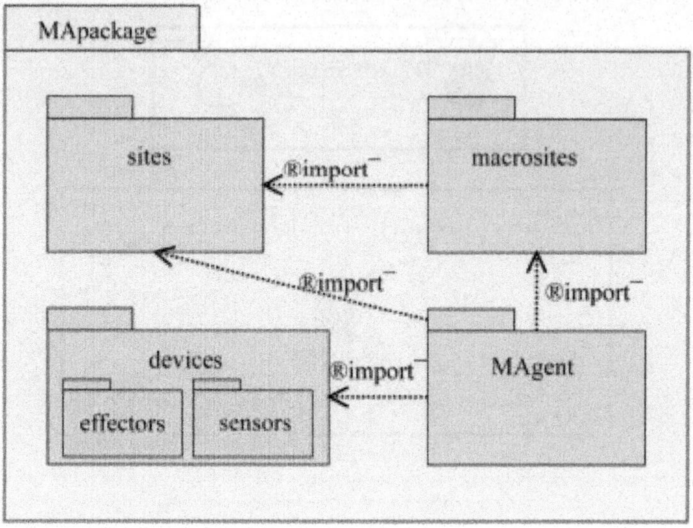

Figure 7: Component organization.

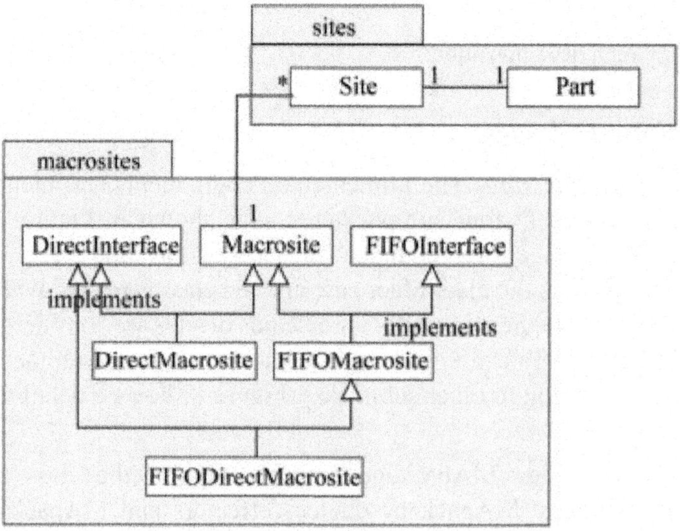

Figure 8: Relationships among sub-packages.

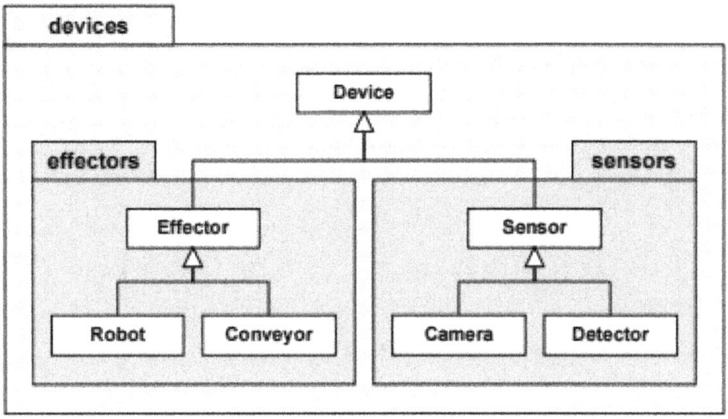

Figure 9: Devices sub-package.

Agent class: The class AgentBase is an abstract class that specializes the class Agent from JADE. This class embeds the behavior of a generic agent that manages all the activities related to a subtask. Every agent must be programmed by extending AgentBase, declaring the knowledge of the corresponding subtask, and instancing the extended subclass. The generic agent provides the facilities for the management of sites and the handling of messages related with the manufacturing task and messages for interacting with other agents; interaction among agents is performed through four kinds of messages regarding information requests and sending about sites, and request/confirmation on the use of sites shared by two or more agents.

Component Identification

The first step for programming an agent is to identify the rules, the sites, the actions, and devices regarding the subtask. For every site in the graph one must declare which agent shares this site, and define if it is a remote site. If a site is used for assembly declare the assembly pattern to follow during the execution of the task. When an in terface site is associated to sensor, only one agent must manage such sensor.

Building the Rule Base

The rules define the strategy of the controller. Every agent has a small set of rules corresponding to the arcs of the sub-graph. Before the writing of the rules it is convenient to enumerate all the information regarding each rule. This may be systematized by the filling of frames, such as shown in Figure 10 for the operation Ident-Loc of the sub-task 1.

The Java coded rules for the agent coordinating the subtask 3 is given below.

```
// implementing rules for the agent named
// ManufacturingAgent3
  void rules()
  {
    // declaring reference to site for place //the
    part
    Site destinationPart = null;
 // GRIP2.TAKE1: Grip2 takes the assembled //part
from ASSB3
    if(assb3.getSite(0).isAssembled() &&
    grip2.isEmpty() &&
    conv2.isEmpty() && destinationPart==null)
    {
         grip2.setContent(assb3.getContent(0));
         assb3.removeContent(0);
         sendUpdateMessages(grip2);
         sendUpdateMessages(assb3);
         //
         destinationPart = conv2;
     }
// GRIP2.TAKE2: Grip2 takes the assembled part
// from ASSB4
    if(assb4.getSite(0).isAssembled()&&
    grip2.isEmpty() &&
    conv3.isEmpty() && destinationPart==null)
```

```
    {
        grip2.setContent(assb4.getContent(0));
        assb4.removeContent(0);
        sendUpdateMessages(grip2);
        sendUpdateMessages(assb4);
        destinationPart = conv3;
    }
// GRIP2.TAKE: Grip2 places the took part in
// CONVin takes the assembled part from ASSB4
    if(!grip2.isEmpty() && destination-
Part.isEmpty() && destinationPart!=null)
    {
  destination-
Part.setContent(grip2.getContent();
        grip2.removeContent();
        sendUpdateMessages(destinationPart);
        sendUpdateMessages(grip2);
        destinationPart = null;
    }
  }
```

Implementation Issues

This method has been demonstrated through the software implementation of several case studies. The software has been written in Java using the JADE framework for supporting the agent definition and task interaction.

The distributed software has been tested on several personal computers interconnected through a local area network; every agent was assigned to single PC. For executing the controller, first the JADE framework is initiated in a computer where an agent may be executed, and then the rest of the agents are started in their computers. For every agent the interaction with a device has been simulated through a visual interface on screen; during a test, users simulate the devices response to commands sent by the controller through the keyboard.

CONCLUSIONS

In this work a method to develop distributed software for manufacturing control systems has been presented. One important issue of the proposed method is the modeling stage; the specifications are transformed into graphical models that contribute to reduce the problems due to ambiguities and incompleteness.

The partitioned task model leads to obtain systematically the knowledge of every agent in the control software, allowing modifying easily the strategies of the subtasks; this feature is useful when task reprogramming is needed. The programming methodology takes advantage of the JADE framework facilities, which permit to the defined agents to be executed into different kinds of platforms.

REFERENCES

1. S. B. Gershwin, "Hierarchical Flow Control: A Framework for Scheduling and Planning Discrete Events in Manufacturing Systems," IEEE Proceedings of S. I. on Discrete Event Systems, Vol. 77, No. 1, January 1989, pp. 195-209.
2. E. López and R. Alami, "A Failure Recovery Scheme for Assembly Workcells," IEEE International Conference on Robotics and Automation, Cincinnati, May 1990, pp 702-707.
3. E. Arjona-Suárez and E. López-Mellado, "A Computer Language for the Modelling of Flexible Manufacturing Systems," Proceedings of the 13th IASTED International Symposium on Robotics and Manufacturing Santa Barbara, California, November 1990, pp. 183-187.
4. C. R. Glassey and S. Adiga, "Conceptual Design of a Software Object Library for Simulation of Semiconductor Manufacturing Systems," Journal of Object Oriented Programing, Vol. 2, No. 4, 1989, pp. 39-43.
5. M. Bakalem, G. Habchi and A. Courtois, "PPS: An Integrated Object Oriented Approach for Modelling and Simulation of Manufacturing Systems," IEEE International Conference on Systems, Man and Cybernetics. Texas, 2-5 October 1994, pp. 2184-2189.
6. G. Ramzi, M. Bakalem and G. Habchi, "An Object Model for Simulation of Manufacturing Systems," IEEE International Conference on Systems, Man and Cybernetics, Vancouver, 22-25 October 1995, pp. 137-142.
7. E. López and E. Medrano, "Object-Based Design of FMS Controllers," Proceedings of IASTED International Conference on Robotics and Manufacturing, Cancun, May 1997. pp. 342-345.
8. N. R. Jennings, K. Sycara and M. Wooldridge, "A Roadmap of Agent Research and Development," International Journal of Autonomous Agents and Multi-Agent Systems, Vol. 1, No. 1, 1998, pp. 7-38. doi:10.1023/A:1010090405266
9. W. Shen and D. H. Norrie, "Agent-Based Systems for Intelligent Manufacturing: A State-of-the-Art Survey," Knowledge and Information Systems, Vol. 1, No. 2, 1999, pp. 129-156.

10. S. A. DeLoach, "Multiagent Systems Engineering: A Methodology and Language for Designing Agent Systems," Agent-Oriented Information Systems'99, Seattle, 1 May 1998, pp. 45-57.
11. S. Bussmann, H. Baumgärtel and M. Klosterberg, "MultiAgent Coordination of Material Flow in a Car Plant," Second International Conference on Practical Applications of Intelligent Agents and Multi-Agent Technology, London, 1997, pp. 227-236.
12. S. Bussmann, "Agent-Oriented Programming of Manufacturing Control Tasks," Proceedings of the 3rd International Conference on Multi-Agent Systems, Paris, 1998, pp. 57-63.
13. D. Ouelhadj, C. Hanachi and B. Bouzouia, "Multi-Agent System for Dynamic Scheduling and Control in Manufacturing Cells," Proceedings of the IEEE International Conference on Robotics and Automation, Leuven, 16-20 May 1998, pp. 2128-2133.
14. J. G. Morales-Montelongo, "Agent-Based Distributed Coordination of Manufacturing Systems," MSC Thesis, Cinvestav Unidad Guadalajara, Mexico, November 2002.
15. F. Bellifemine, A. Poggi and G. Rimassa, "JADE—A FIPA—Compliant Agent Framework," Proceedings of International Conference on Practical Applications of Intelligent Agents and Multi-Agent Technology, London, April 1999, pp. 97-108.
16. "Foundation for Intelligent Physical Agents," Specifications, 2005. http://www.fipa.org.

Chapter 2

RELIABILITY MODELING AND OPTIMIZATION STRATEGY FOR MANUFACTURING SYSTEM BASED ON RQR CHAIN

Yihai He[1,2], Zhenzhen He[1], Linbo Wang[1], and Changchao Gu[1]

[1]School of Reliability and Systems Engineering, Beihang University, Beijing 100191, China

[2]Department of Systems Engineering and Engineering Management, City University of Hong Kong, Kowloon, Hong Kong

ABSTRACT

Accurate and dynamic reliability modeling for the running manufacturing system is the prerequisite to implement preventive maintenance. However, existing studies could not output the reliability value in real time because their abandonment of the quality inspection data originated in the operation process of manufacturing system. Therefore, this paper presents an approach to model the manufacturing system reliability dynamically based on their operation data of process quality and output data of product reliability. Firstly, on the basis of importance explanation of the quality variations in manufacturing process as the linkage for the manufacturing system reliability and product inherent reliability, the RQR chain which could represent the relationships between them is put forward, and the product qualified probability is proposed to quantify the impacts of quality variation in manufacturing process on the reliability of manufacturing system further. Secondly, the impact of qualified probability on the product inherent reliability is expounded, and the modeling approach of manufacturing system reliability based on the qualified probability is presented. Thirdly, the preventive maintenance optimization strategy for manufacturing system driven by the loss of manufacturing quality variation is proposed. Finally, the validity of the proposed approach is verified by the reliability analysis and optimization example of engine cover manufacturing system.

INTRODUCTION

To meet the demands of high reliability and long life of the product, integrated analysis, assurance, and optimization for reliability are required to be carried out in the lifecycle of design, manufacture, and usage. However, for a long time, most of traditional reliability studies had merely focused on the design and usage stages, and reliability technology suitable for the manufacturing process has always been ignored, lacking proper attention it deserved, which caused the serious degradation of product reliability after batch production frequently, and resulting in a high infant failure rate, and the product inherent reliability could not meet the increasingly stringent design reliability requirements [1]. As we all know, product is the output of the manufacturing process which is the implementation form of the manufacturing system. Therefore, the reliability of final produced product is closely related to the reliability of manufacturing system and the quality of manufacturing process. Usually, even a good design cannot guarantee that the manufactured products achieve the satisfactory reliability when the design quality of manufacturing system is poor [2]. Thus, it can be seen that integrating the reliability modeling and optimization for manufacturing system is crucial to ensure the product reliability.

Practices have proved that uncertain factors like quality variation in manufacturing process and deteriorations of system components could lead to the degradation of manufacturing system, which should affect the quality of manufacturing process interactively. And when quality variations are cumulated and amplified, the number of potential defects of products is arising, which would trigger the decline of product inherent reliability finally. In order to minimize the decline of product inherent reliability with respect to the design specifications, identifying and optimizing the critical factors in manufacturing that contribute to the product reliability degradation systematically are becoming the research focus of reliability engineering currently, and how to carry out product reliability oriented reliability modeling and optimization of manufacturing system is the most urgent and task.

At different nodes of product life cycle, product reliability exhibits different characteristics. Murthy [3] defined the evolution chain which transfers product reliability from design, manufacturing, transportation, sale, and usage, enriching notation of product reliability at different stages, and named the reliability in manufacturing as product inherent reliability. Then, Jiang and Murthy [4] pointed out the negative impact of variations on the reliability during the product life cycle via the transmission chain and pointed out that the quality variations and assembly errors are the root reasons causing the deterioration of product inherent reliability. As to product inherent reliability in manufacturing, Li et al. [5, 6] noted that both the reliability of manufacturing system and quality of manufacturing process are the critical roles to ensure and improve the product quality and reliability. Inman et al. [7] believed that performance of manufacturing system severely restricted product quality and reliability, and upgrading the manufacturing equipment or adjusting the technological process could promote the ability of manufacturing system as well as ensuring and optimizing product quality and reliability. To some extent, the ability of trouble free operation of manufacturing system determines the level of inherent reliability formed in manufacturing process.

Traditional reliability modeling of manufacturing system tends to follow the classic reliability block diagram method, fault tree analysis, Petri nets, and so forth, which caused a comprehensive analysis and dynamical assessment for manufacturing system to be complex or inconvenient. Based on the data of system operation and maintenance, Li and Ni [8] used the maximum likelihood estimation method to estimate the reliability of manufacturing system, which provided the basis for carrying out preventive maintenance of manufacturing system. Lin and Chang [9] proposed the limited manufacturing network model, and after mining operating failures and rework data, an analysis model of manufacturing system reliability was established. Li et al. [10] created a prediction model of manufacturing system using the grey model, and the author stated that the weaknesses of manufacturing system could be identified by the proposed model. Considering the plenty of quality data existing in manufacturing process, Chen and Jin [11, 12] put forward a Quality-Reliability chain model based on the interaction between manufacturing process quality and manufacturing system reliability, and the reliability analysis and maintenance optimization of manufacturing system were expounded based on the proposed Quality-Reliability chain. Zhang et al. [13] presented a reliability modeling approach of manufacturing system using dimensions of process quality. Rafiee et al. [14] analyzed four typical vibration modes and their effects on the degradation rate of manufacturing process and modeled the complex manufacturing system reliability like MEMS and so on. Regarding the maintenance strategy of manufacturing systems, Li et al. [15] investigated the economic production quantity model jointly considering product deterioration and proposed an EPQ (economical production quantity) model for deteriorating production system and items with rework. Gong et al. [16] explored an adaptive maintenance model of the process environment to diagnose the progressive faults in manufacturing systems. Tlili et al. [17] proposed a new modeling approach based on the fact that the degradation process is modeled by the wiener process. Hajej et al. [18–21] studied integrated maintenance strategies and policies jointly considering the optimization problems of subcontracting, product returns, lease contract, and so forth, which provide a solid foundation to develop the integrated maintenance strategies optimization for manufacturing system. Mifdal et al. [22] presented a joint optimization approach of maintenance and production planning for a multiple-product manufacturing system, which could establish sequentially an economical production plan and an optimal maintenance strategy considering the influence of the production rate on the system's degradation.

As can be seen from the above literature analysis, the interaction between product inherent reliability and manufacturing system reliability is not defined, and studies on the product reliability oriented reliability modeling and optimization of manufacturing system are few, which are in urgent need to develop reliability assurance in manufacturing. Therefore, in order to promote the joint reliability optimization of manufacturing system and the produced product, an integrated model named chain is proposed by extending the Quality-Reliability chain in this paper, the chain could describe the coeffects of the manufacturing system reliability , manufacturing process quality , and product inherent reliability , and the impact of manufacturing system reliability on the product inherent reliability is expounded specifically based on product qualified probability. At the same time, the quantitative analysis model and optimization strategies of manufacturing system reliability are given. Comparing to previous related studies in the frame of integrated reliability and maintenance optimization of manufacturing system, the main contributions of this paper are as follows:

As can be seen from the above literature analysis, the interaction between product inherent reliability and manufacturing system reliability is not defined, and studies on the product reliability oriented reliability modeling and optimization of manufacturing system are few, which are in urgent need to develop reliability assurance in manufacturing. Therefore, in order to promote the joint reliability optimization of manufacturing system and the produced product, an integrated model named RQR chain is proposed by extending the Quality-Reliability chain in this paper, the RQR chain could describe the coeffects of the manufacturing system reliability R_m, manufacturing process quality Q_p, and product inherent reliability R_p, and the impact of manufacturing system reliability on the product inherent reliability is expounded specifically based on product qualified probability. At the same time, the quantitative analysis model and optimization strategies of manufacturing system reliability are given. Comparing to previous related studies in the frame of integrated reliability and maintenance optimization of manufacturing system, the main contributions of this paper are as follows:

- RQR chain is brought forth and the product qualified probability is proposed to quantify the impacts of manufacturing process quality on the manufacturing system reliability for the first time. The product qualified probability driven RQR chain could make integrated reliability optimization of manufacturing system and produced product possible.

- A reliability modeling approach of manufacturing system based on the proposed qualified probability is presented. The impact of qualified probability on the inherent reliability of produced product oriented is expounded at the first time in this paper, and the concept of the reliability of manufacturing is extended by including the requirement of product quality that is qualified by time t in this proposed approach.
- The preventive maintenance optimization strategy for manufacturing system driven by the loss of manufacturing quality variation is proposed by the aid of the product qualified probability. The proposed product qualified probability based optimization approach could make the real time preventive maintenance possible when the abnormal quality variations occurred in the produced work pieces in the batch production.

The rest of the paper is organized as follows. Section 2 emphasizes the role of manufacturing process quality variations as a transfer bridge for analyzing the influence of manufacturing system reliability on product inherent reliability, and the RQR chain based on product qualified probability is put forward. Modeling of manufacturing system reliability based on the qualified probability is analyzed in Section 3. With reference to the results of the proposed model, Section 4 presents some optimization strategies driven by product quality loss. Section 5 discusses the application mode and the effects of the proposed method in an automotive cylinder head manufacturing system. Finally, conclusions are drawn in Section 6.

QUALITY ORIENTED RQR CHAIN

RQR Chain

In the manufacturing process, normal and abnormal variations from man, machine, material, method, measurement, and environment (5M1E) should be accumulated and inherited by the work pieces, resulting in the variations of product dimensions eventually. Variations of product quality formed in the manufacturing process are the basic factors influencing the product inherent reliability. As manufacturing system acts as the material carrier of manufacturing process, its stability determines the quality of the manufactured product. Thus, it can be concluded to a great extent that the product inherent reliability is relying on the reliability of manufacturing system with a fixed product design scheme, product reliability in manufacturing process would be deteriorated by those abnormal factors like wear degradation or failures of system components or some assembly errors, and so forth, and the number of potential quality defects lying in the manufactured products should increase successively, resulting in the product inherent reliability failing to satisfy the design requirements. It is obvious that if the reliability of manufacturing system cannot be guaranteed effectively, the dimensional parameters of the manufactured work pieces will be deteriorated constantly (such as abnormal dimensional variations in the manufacturing process), and contrasted with design reliability, these variations would bring about a decline of product inherent reliability. To make matters worse, products composed by these unreliable components are prone to some unpredictable fatal failures. In order to describe the relationship among manufacturing system reliability R_m, process quality Q_p, and product inherent reliability R_p, a conceptual model of RQR chain is put forward and shown in Figure 1. As shown in Figure 1, the product quality in manufacturing process is the linkage of the manufacturing system reliability and product inherent reliability which is clearly reflected by the RQR chain, and in the view of the mathematical modeling, the RQR chain could be divided into the following three layers from the top-down: physical layer, parameter layer, and data layer. The three-tier construction of RQR chain could enable the integrated analysis for the reliability of manufacturing system based on the quality data in manufacturing and reliability data in field as much as possible.

- Physical Layer of RQR Chain. The coeffects of manufacturing system, manufacturing process, and product are described clearly in this layer. Specifically, due to system components wear or failures caused by friction wear, reliability of manufacturing system R_m gets lower. During the execution of functional requirements from manufacturing system, quality in manufacturing process Q_p results in abnormal variations accompanied by degradation of the manufacturing system. Correspondingly, owing to these quality variations, there may be potential defects retained in the final product, which has a negative effect on product reliability R_p. That is to say, reliability of manufacturing system determines directly the stability of manufacturing process quality, and then the reliability of produced product is subjected to the stability of manufacturing process quality.
- Parameter Layer of RQR Chain. The indicating parameters of the mentioned coeffects in RQR chain are given in this layer, the quantitative indicators of manufacturing system reliability, product qualified probability, and product inherent reliability are presented, and these parameters are the carriers of the coeffects in RQR chain.

- Data Layer of RQR Chain. The product key quality characteristics including key control characteristics and are presented in the data layer, which provides the data source to compute the value of parameters given in the parameter layer. As shown in Figure 1, the quality of manufacturing process Q_p is the core of the RQR chain, which links manufacturing system reliability R_m and product reliability R_p. To accurately analyze manufacturing system reliability oriented by product inherent reliability and further optimize the manufacturing system, quantifying the quality of manufacturing process is bound to the prerequisite. Typically, current studies adopt product qualified rate to characterize the quality of manufacturing process, which simply reflects the cursory state of process quality Q_p in the form of scalar quantity. And it is still actually inconvenient to carry out system reliability R_m modeling and optimization that considers process quality variation information in the form of vector. Therefore, based on the definition of product quality, the big data like variation vector of manufacturing process and system components degradation is combined firstly to put forward product qualified probability to represent manufacturing process quality in vector.

Product Qualified Probability

Product qualified rate is often used in traditional quality control to describe the quality of manufactured products and the capability of manufacturing process. It is deemed that if all the key quality characteristics of products are within their tolerance limits at time t, products are thought to pass through the inspection and there is no substandard product throughout the manufacturing process. Count the number of qualified products and figure out how much it accounts for the total inspected products, and, namely, the product qualified rate is obtained. Here, product qualified rate integrates count information of a single product as a whole and characterizes the quality of manufactured products in the form of scalar. However, the values of the detected quality characteristics are obtained in the vector form with the advancement of detection technology in practical engineering applications usually.

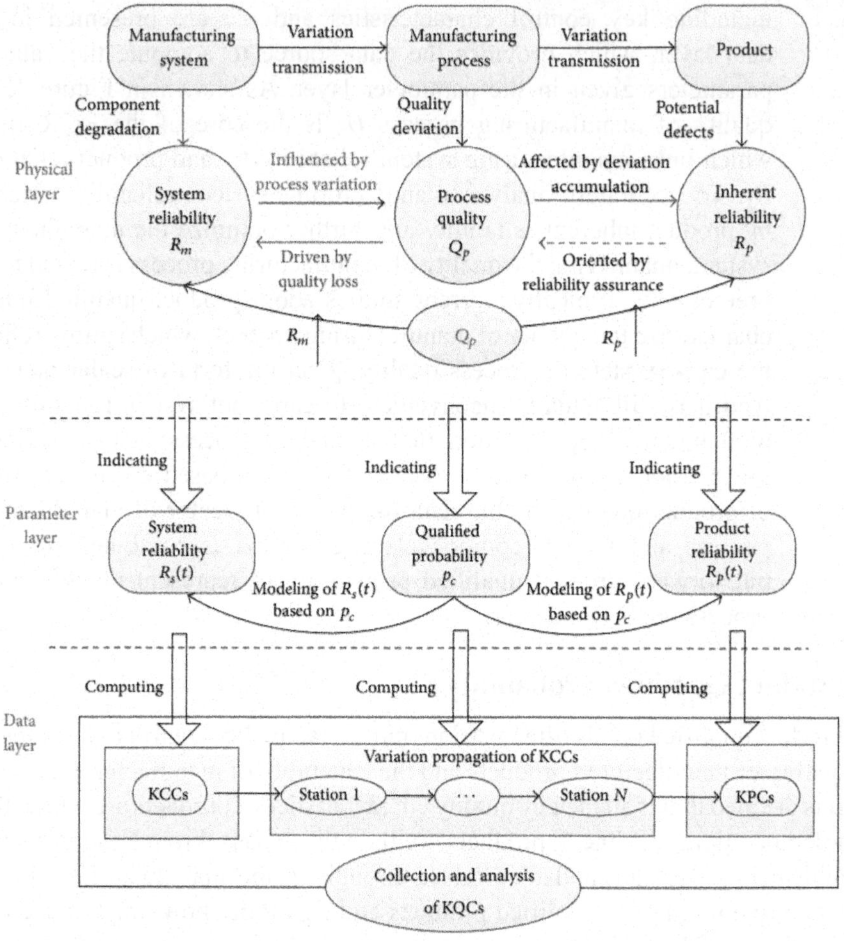

Figure 1: Conceptual model of RQR chain.

Different from the scalar expression of product qualified rate, defining quality levels and grades often requires quantifying the extent of how values of product key quality characteristics approximate the predetermined targets further. And thus, measurable information such as parameters of different components inside a single product should be the concern of quality control. At present the qualified rate is basically the only evidence to determine whether a batch product is qualified or not, and the judgment is arbitrary, which will result in neglecting and omitting useful variation information of those measurable key quality characteristics under modern vectorial measurement environment. And it turns out to be not conducive to fully exploit and utilize process quality data and then carry out a comprehensive analysis of both product reliability and manufacturing system reliability. Accordingly, product qualified probability oriented by quality grade is proposed and the vector space of different parameters that constitute product key quality characteristics is built in this paper, and thus vectorization of qualified degree of products is realized. When values of the key quality characteristics are closer to the predetermined ones, both quality grade and the corresponding degree of qualified product become higher. Namely, it corresponds to a higher product qualified probability. From the perspective of population and sample, Figure 2 contrasts the differences and relations between traditional qualified rate and the proposed qualified probability based on the interactive scalar of the whole population and the vector of the single sample.

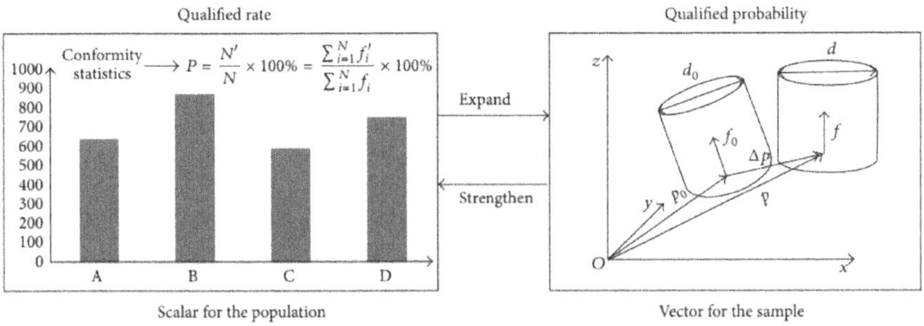

Figure 2: Comparison of qualified rate and qualified probability.

Given the threshold value θ_k of key quality characteristics, product qualified probability can be expressed as

$$P_k^c = \Pr\{\Delta y(k) \le \theta_k\}. \tag{1}$$

Here, $\Delta(k)$ denotes the variation of the actual characteristic dimension to the target value at station k, and it can be obtained through the SoV [23] (Stream of Variation) theory. Generally, the threshold value θ_k can be acquired by means of product quality specifications and the popular process capability index. Set the tolerance range of products to be USL (upper specification limit) and LSL (lower specification limit), with process capability index being C_{pk} (actual process capability) and variance of key quality characteristics being θk, and their values should be calculated via the formula $C_{pk} = (\text{USL} - \text{LSL})/6\sqrt{\theta_k}$, $\theta_k = [6C_{pk}/(\text{USL} - \text{LSL})]^2$.

of Set $f(x) = \Delta y(k) - \theta_k$, and ignore the noise of manufacturing process and the measurement. Since (x) is a nonlinear function of $(k) = [x_1, x_2, \ldots, x_n]^T$, for simplicity, Taylor expansion is used to linearize $f(x)$ as below:

$$f(x) = a_0 + \sum_{i=1}^{n} a_i x_i \quad (a_i \text{ is constant}),$$

$$\beta = \frac{\mu_f}{\sigma_f} \tag{2}$$

$$= -\frac{a_0 + \sum_{i=1}^{n} a_i \mu_{x_i}}{\sqrt{\sum_{i=1}^{n} a_i^2 \sigma_i^2 + \sum_{i=1}^{n} \sum_{\substack{j=1 \\ j \neq i}}^{n} a_i a_j \mathrm{Cov}\left(x_i, x_j\right)}}$$

$$\left(\mu_f = a_0 + \sum_{i=1}^{n} a_i \mu_{x_i}\right) \tag{3}$$

As $\mu_x = (\mu_{x_1}, \mu_{x_2}, \ldots, \mu_{x_n})$ is the mean point of independent variables of $f(x)$, correspondingly, mean of the dependent variable is denoted by $\mu_f = f(\mu_{x_1}, \mu_{x_2}, \ldots, \mu_{x_n})$ and the standard variation by $\sigma_f^2 = \sum_{i=1}^{n} a_i^2 \sigma_i^2 + \sum_{i=1}^{n} \sum_{\substack{j=1 \\ j \neq i}}^{n} a_i a_j \mathrm{Cov}(x_i, x_j)$. When process variables obey iid (independent and identically distributed), there exists $\sigma_f^2 = \sum_{i=1}^{n} a_i^2 x_i^2$. And product qualified probability of manufacturing process should be derived as

$$P_k^c = \Pr\{\Delta y(k) \leq \theta_k\} = \Pr\{f(x) \leq 0\}$$

$$= p\left\{\frac{f(x) - \mu_f}{\sigma_f} < -\frac{\mu_f}{\sigma_f}\right\} = \Phi(-\beta) \tag{4}$$

RELIABILITY MODELING OF MANUFACTURING SYSTEM BASED ON RQR CHAIN

Analysis of Product Inherent Reliability Oriented by Qualified Probability

Manufacturing processes are comprised of raw materials purchasing, parts machining, components assembling, and performance testing. With reference to reliability requirements of the design phase, parts that are either outsourced or homemade are operated by machining, assembling, testing, and so forth, and thus the Work in Process (WIP) or the end products are finally produced. Influenced by a variety of variation factors, the product inherent reliability gets lower than requirements of design reliability usually. Jiang and Murthy [4] noted that nonconforming components and assembly errors would have undesirable effect on the manufactured product reliability, and the nonconforming components are the basic adverse factors. Based on this standpoint, the product qualified probability is brought forward to measure the potential influence of nonconforming components on the product inherent reliability.

Assuming that failure distribution of products in the design phase is $F_0(t)$, the design reliability of products $R_o(t)$ can be expressed as $R_o(t) = 1 - F_0(t)$. Correspondingly, failure density function and failure rate function are denoted by $f_0(t) = dF_0(t)/dt$ and $r_0(t) = f_0(t)/R_o(t)$, respectively. Similarly in the manufacturing process, failure distribution of nonconforming components is (t) with reliability function being $R_h(t) = 1 - H(t)$, failure density function being $h(t) = dH(t)/dt$, and failure rate function being $r_h(t) = h(t)/R_h(t)$. According to the number of the nonconforming components and the unqualified degree of the nonconforming components, define the probability that results in nonconforming components as q and the qualified probability of components as p_c.

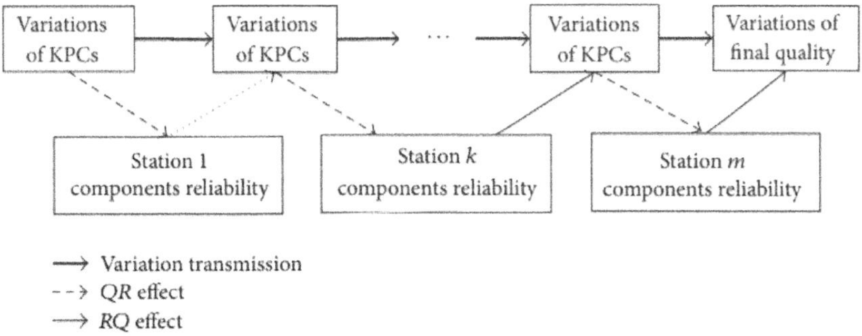

Figure 3: Interaction between quality variations and components reliability in multi-station process.

Then with all the components being fully qualified, namely, $p_c = 1$, it is the only case of occurrence of nonconforming components that affects the reliability of manufactured products, which is denoted by

$$R_1(t) = (1 - q) R_o(t) + qR_h(t). \tag{5}$$

Obviously, with q being equivalent to 0, the manufactured reliability equals the design reliability, indicating that no nonconforming components occur, and with q being equivalent to 1, it happens to be the opposite from the former case of q=0, indicating that all components fail to be conforming in manufacturing process.

When it comes to components being not fully qualified with $0 < p_c < 1$, the manufactured reliability is subjected to the occurrence of nonconforming components and the related unqualified level jointly, and, then, inherent reliability of the manufactured product is as follows:

$$R_2(t) = (1-q)(1-p_c)R_o(t) + qp_cR_h(t). \tag{6}$$

It is evident that with $p_c = 1$, $R_2(t) = qR_h(t) \leq R_1(t) = (1-q)R_o(t) + qR_h(t)$, which shows that integration of both the occurrence of nonconforming components and their unqualified level is conducive to a more precise estimation of the manufactured reliability. While simply based on the case whether components are unqualified, inherent reliability of the manufactured products will be blindly overestimated, which would have a great negative impact on the objective analysis and optimization for the reliability for manufacturing system.

Modeling Reliability of Manufacturing System Based on Qualified Probability

Reliability of the manufacturing system is an important factor to ensure product quality and productivity. Considering the difficulty in quantifying the reliability issues timely caused by product dimension variations, usually those reliability issues are easily neglected in system design and optimization. And unfortunately, with many failures or performance degradation failing to be diagnosed in real time, further identification or predication is of nonsense, which may result in unnecessary downtime for machines and reduce the production efficiency. Mechanisms of performance degradation and occurrence of failures of system components are versatile and complicated. Both components themselves of this current station and quality variations from the upstream station make a difference to the performance of components as well as the manufacturing system. Reliability analysis of multistation manufacturing system notes that via the transmission of quality variations, interaction between variations of key product characteristics (KPCs) and reliability of system components is also transferred through the stations. Thus, the statistical correlation between manufacturing system component failures and unqualified products could be shown in Figure 3. Accordingly, we are badly in need for a new connotation of reliability of manufacturing system. Degree of reliability is generally used to measure the reliability of the manufacturing system which refers to the probability a system completes its intended functions under specified conditions within the specified time. Not only the expected functions of a manufacturing process should consider the uptime of manufacturing system, but also the quality of manufactured product should also be involved to holistically assess the performance and reliability of manufacturing system. As a conclusion, the reliability of manufacturing system (t) should be identified as the probability that system components do not fail themselves and at the same time manufactured products are completely qualified within a period of time, which can be expressed as

$$R(t) = P\{\text{system does not fail by time } t \cap \text{product quality is qualified by time } t\}. \tag{7}$$

Define R^F and R^Q to represent cases where system does not suffer catastrophic failures and product quality is qualified by time t, respectively. And (t_k) means the performance state of system components at the endpoint of t_k. To sum up, the reliability of manufacturing system could be rewritten as

$$R(t) = P\{\text{System does not fail by time } t \cap \text{Product quality is qualified by time } t\} = \Pr\{R^F \cap R^Q\} = \Pr\{R^F \mid Z(t)\} \cdot \Pr\{R^Q \mid Z(t)\}. \tag{8}$$

Here, $\Pr\{R^Q \mid Z(t)\}$ corresponds to the product qualified probability mentioned in Section 2.2. Namely, there exists $\Pr\{R^Q \mid Z(t)\} = P_k^c = \Phi(-\beta)$. So, the process of how to determine the $\Pr\{R^F \mid Z(t)\}$ will be highlighted in the following.

In general, the performance state and relative operating conditions of system components determine the reliability of manufacturing system. Either components failures or degradation by wear could cause a decline in reliability of manufacturing system. Since status information like wear or degradation of tools comes along the running of manufacturing system, failure data may not occur necessarily with the advancement of manufacturing technologies. Therefore, this paper prefers to consider information of wear or degradation of system components as the main factors affecting the reliability of manufacturing system. The wear and tear of system components are accumulated and increased along with the front and back work stations one by one. Let $\Delta(k)$ be the amount of wear from the individual station k, and the cumulative amount of wear by station k is $Z(k) = Z(k-1) + \Delta(k)$. If all the wear or degradation processes are independent and identically distributed as most mechanical products, according to the central limit theorem, (k) can be rewritten as

$$Z(k) = \sum_{j=1}^{k} \Delta(j) \approx N(k \cdot E[\Delta(j)], k \cdot \text{Var}[\Delta(j)]). \tag{9}$$

Define (t) as the probability that system component fails at station k+1 while it still functions well at station k by time t, namely, the failure rate of the system component. Assuming that failure of the individual component is subjected to an exponential distribution (for high reliability of complex systems, exponential distribution can approximately model the failure distribution for those system components), reliability of system components at station k is expressed as (excluding the impact of component wear on component strength, the failure rate can be regarded as a constant):

$$R_k^F(t) = e^{-\lambda_k(t) \cdot t}. \tag{10}$$

When the input products have problems of quality variations, wear and tear of system components will be accelerated as

$$\lambda_k(t) = \lambda_0(t) + E\left(\alpha_k (X(k) - m_k)^2\right). \tag{11}$$

Here, $\lambda_0(t)$ is the initial failure rate irrespective of impact of input product quality on system components; $X(k)$ is the practical quality index at station k; m_k stands for the standard value of $X(k)$ with α_k being the correction coefficient which reflects the impact of input quality variation on wear of components. And the reliability of the whole system could be presented as

$$R^F(t) = \prod_{i=1}^{n} R_i^F(t). \tag{12}$$

Suppose that η is the allowable maximum amount of components wear, ε_k is the degradation coefficient of relevant performance, w_0 is the initial rate of degradation which corresponds to time $t_k = 0$, and the probability that no failures occurred for the whole system by time t is as follows:

$$\Pr\left\{R^F \mid Z(t)\right\} = P\left(\prod_{i=1}^{n} R_i^F \mid Z(t) < \eta\right)$$

$$= P\left(\prod_{i=1}^{n} e^{-[\lambda_{0i}(t)+E(\alpha_i(X(i)-m_i)^2)]\cdot t_i} \mid \sum_{k=0}^{h} w(t) < \eta\right)$$

$$= P\left(\prod_{i=1}^{n} e^{-[\lambda_{0i}(t)+E(\alpha_i(X(i)-m_i)^2)]\cdot t_i} \mid h\left(w_0 + e^{-\varepsilon_k \cdot t}\right)\right.$$

$$\left. < \eta\right)$$

$$= \prod_{i=1}^{n} \exp\left(-\left[\lambda_{0i}(t) + E\left[\alpha_i (X(i) - m_i)^2\right]\right] \cdot t_i\right)$$

$$\cdot \int_0^{\eta} \exp\left(-h\left(w_0 + e^{-\varepsilon_k \cdot t}\right)\right) d\varepsilon. \tag{13}$$

Combined with formula (8), the final expression of the reliability of manufacturing system based on the product qualified probability (t) could be written as

$$R(t) = \Pr\left\{R^F \mid Z(t)\right\} \times \Pr\left\{R^Q \mid Z(t)\right\}$$

$$= \exp\left(\sum_{i=1}^{n}\left(-\left[\lambda_0(t) + E\left[\alpha_i (X(i) - m_i)^2\right]\right] \cdot t\right)\right)$$

$$\cdot \int_0^{\eta} \exp\left(-h\left(w_0 + e^{-\varepsilon_k \cdot t}\right)\right) d\varepsilon_k$$

$$\cdot \Phi\left(-\frac{\alpha_0 + \sum_{i=1}^{n} \alpha_i \mu_{x_i}}{\sqrt{\sum_{i=1}^{n} \alpha_i^2 + \sum_{i=1}^{n} \sum_{\substack{j=1 \\ j\neq i}}^{n} \alpha_i \alpha_i \mathrm{Cov}\left(x_i, x_j\right)}}\right)$$

$$= \exp\left(\sum_{i=1}^{n}\left(-\left[\lambda_0(t) + E\left[\alpha_i (X(i) - m_i)^2\right]\right] \cdot t\right)\right)$$

$$\cdot \int_0^{\eta} \exp\left(-h\left(w_0 + e^{-\varepsilon_k \cdot t}\right)\right) d\varepsilon_k \Phi(-\beta). \tag{14}$$

According to formula (14), if the interaction between product quality and components failures or performance degradation has been ignored, the reliability of manufacturing system should be overestimated, which would endanger the quality and reliability of manufactured products, and the final goal to obtain high reliable products cannot be fulfilled. The correlation modeling of product quality and components reliability should help us to establish a more objective and accurate model, which makes a more authentic and practical assessment of the reliability of manufacturing system, and provides specific goals and directions for further improving the reliability of manufacturing system.

PREVENTIVE OPTIMIZATION STRATEGY FOR RELIABILITY OF MANUFACTURING SYSTEM DRIVEN BY QUALITY LOSS IN MANUFACTURING PROCESS

Requirements Analysis of Dynamic Optimization

The purposes of conducting analysis, modeling, and assessment of the reliability manufacturing system are mainly designed to guide the timely maintenance for failed components or degraded ones, which would help to improve the quality level of the manufactured products, and how to obtain the optimal strategy in real time of the reliability of manufacturing system has long been a mathematical puzzle in industrial engineering [16, 17]. The previously proposed reliability model of manufacturing system highlights the role of product quality variation and its interaction effect with components reliability. Therefore, not only maintenance cost of manufacturing system itself but also the quality loss in manufacturing process should be simultaneously considered when developing reliability optimizing strategies for manufacturing system. That is to say, the total costs for optimizing manufacturing system should include quality loss caused by the variations that occurred in manufacturing process, maintenance costs by component failures, and so forth. According to the reliability model presented in Section 3.2, in order to realize dynamic or real time optimization, the product quality loss should be taken into account to formulate the optimization strategies. For the convenience of the computation, the assumptions and hypotheses are given firstly as below

- Suppose the maintenance cycle is T, and $T = k\Delta t$, $k \in N$, where k is a constant. Then the extent of performance degradation for each component can be determined. Denote the degradation state of component i by z_i ($i = 1, 2, \ldots, n$) and assume that the cost needed for conducting a state inspection for component i is C_i, where the inspection time can be neglected. Without loss of generality, Δt can be considered as 1, and thus T can be directly written as k in the context. Since the operating time for each component may be different from each other, time for state inspection separately may be not necessarily synchronized.

- Assume there are two ways of maintenance of each component, which includes the postmaintenance and preventive maintenance, and inspection time of the maintenance is fixed at t_k. If the maintenance is assumed to be an overhaul or a complete replacement, functions of components are believed to be fully restored by repairment. Define costs for postmaintenance and preventive maintenance as C_c^i and C_p^i, respectively, where the magnitude relation conforms to $C_c^i > C_p^i$
- To quantify the impact of components degradation on product quality in manufacturing process, Taguchi function is adopted to analyze the quality loss caused by components degradation or error propagation, and so forth.

Optimization Strategy Decision-Making Model Driven by Quality Loss

Based on the above analysis and assumptions, the objective of optimization strategy for the reliability of manufacturing system is to minimize the costs of quality loss, postmaintenance, and preventive maintenance simultaneously. And it can be represented as a constraint optimization problem in the following expression:

$$\text{Min EC} = \text{Min } f\left(C_m, C_q\right), \tag{15}$$

where EC represents the expected cost of the optimization strategy, C_m means the expectation of average maintenance costs comprised of postmaintenance costs and preventive maintenance costs, and C_q signifies the expectation of average quality loss.

It is generally believed that failure rate of each system component is fixed to the same as λi. And to be consistent with the prementioned assumption in Section 3.2, the failure of the individual component is subjected to an exponential distribution, and the reliability of system components at station i is expressed as $e^{-\lambda_i \cdot t}$. Accordingly, the preventive maintenance at the unit expense of C_p^i for the reliable system component gets the cost of $C_p^i e^{-\lambda_i t}$, whereas the postcorrective maintenance at the unit expense of C_c^i for the unreliable system component gets the cost of $C_c^i(1 - e^{-\lambda_i t})$. And thus, for one single system component by each inspection time, the total maintenance cost generally consists of the two mentioned parts of the preventive $C_p^i e^{-\lambda_i t}$ and the postcorrective $C_c^i(1-e^{-\lambda_i t})$.

What is more, with lifetime of component i being s_i and the number of these system components being n, the average maintenance costs C_m can be extended as

$$C_m = \sum_{i=1}^{n} \frac{C_p^i e^{-\lambda_i t} + C_c^i \left(1 - e^{-\lambda_i t}\right)}{\int_0^{s_i} e^{-\lambda_i t} dt}. \tag{16}$$

And it can be simplified as the following expression:

$$C_m = \sum_{i=1}^{n} \lambda_i \left[\frac{C_p^i}{1 - e^{-\lambda_i s_i}} + C_p^i - C_c^i \right]. \tag{17}$$

And the quality loss function based on the Taguchi function is expressed as

$$L(k) = q\left(X(k) - m_k\right)^2. \tag{18}$$

Here, (k) is the quality loss with q the constant responding to the coefficient of quality loss, $X(k)$ representing product key quality characteristic, and m_k being the specification value for $X(k)$. Take the mathematical expectation for formula (18) as

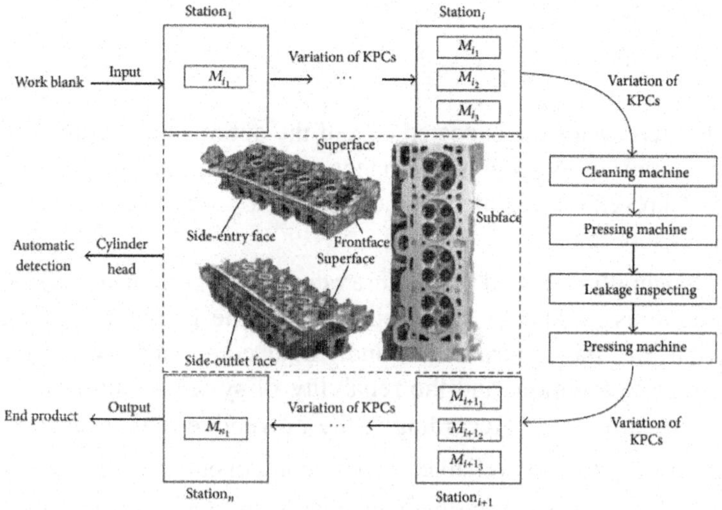

Figure 4: An illustration of key manufacturing process for cylinder head manufacturing system

$$E\left[L\left(k\right)\right] = q\mathrm{Var}\left[X\left(k\right)\right] + q\left[E\left(X\left(k\right)\right) - m_k\right]^2, \tag{19}$$

wherein $E[X(k)]$ and $\mathrm{Var}[X(k)]$ are the mathematical expectation and variance of $X(k)$, respectively.

When quality loss is taken into account of the optimization costs, estimation for the costs will be more conservative actually. After applying the exponential distribution to describe cost of quality loss by time t, the expectation of average quality loss C_q should be rewritten as

$$C_q\left(t\right) = C_0 \exp\left\{-E\left[L\left(k\right)\right] \cdot t\right\}, \tag{20}$$

Where C_0 is the initial cost, meaning there is no loss of product quality. Integrating formula (16) and formula (20) into formula (15), the optimization objective is converted to minimize both maintenance costs and product quality loss, and the final formula is as follows:

$$\mathrm{Min\,EC} = \mathrm{Min}\, f\left(\sum_{i=1}^{n} \lambda_i \left[\frac{C_p^i}{1 - e^{-\lambda_i s_i}} + C_p^i - C_m^i\right], \right.$$

$$\left. \lim_{T \to \infty} \frac{\int_0^T C_0 \exp\left\{-E\left[L\left(k\right)\right] \cdot t\right\} dt}{T}\right). \tag{21}$$

CASE STUDY

Background

As a key part of the engine, the cylinder head is mounted on the upper end of the cylinder block with cylinder head bolts, forming a sealed combustion chamber together with the cylinder block. Coordinating with components of intake and exhaust valves, fuel injectors, pneumatic valves and others, the cylinder head plays a vital role in controlling fully combustion of the air and fuel inside the cylinder. And thus, the system and process for manufacturing the cylinder head turns to be particularly complex and elaborated, key quality characteristics like the surface roughness, geometrical shape, machining dimensions and location precision are needed special attention and monitoring. Usually, variations of dimensions have a great impact on the assembly precision and even the overall performance of the engine. The cylinder head is comprised of the following six components: a superface, a subface, a frontface, a backface, a side-entry face, and a side-outlet face. The machining features are reflected in the complicated structures of surfaces and holes. Accordingly, how to assure the machined surfaces and holes with high precision is the core function of the manufacturing system of cylinder heads, which includes function modules of cutting, clamping, controlling, testing, and clearing.

The key operation processes of the cylinder head manufacturing system are shown in Figure 4. In this paper, key design characteristics of the cylinder head machined by the studied manufacturing system are listed in Table 1. In practical manufacturing process, due to the integrated effects of various variations, reliability of the cylinder head fails to meet the designed reliability requirements, resulting in a phenomenon of reliability degradation relative to the design reliability and an unsatisfactory response from customers. With mechanical inspection highly automated, it is urgent to reduce the interference from the vector space of key product characteristics and further to monitor the cylinder head quality effectively throughout the entire manufacturing process. Typically, optimization of manufacturing system reliability is the fundamental premise to control the cylinder head quality and reliability. However, whether to replace worn tools or conduct regular maintenance for a high level of system reliability, the level of qualified rate and reliability of the produced products still remained low. How to conduct the reliability of manufactured product oriented modeling and optimizing of manufacturing system reliability are what the proposed RQR chain tries to contribute.

Table 1: Key design characteristics of the cylinder head.

Basic dimension	Machining equipment	Process description	Major characteristics	Design specification
B_1	Machining center	Drilling hole A	Diameter	$\phi 13_{0}^{+0.018}$
B_2	Machining center	Drilling hole B	Diameter	$\phi 12.5_{0}^{+0.018}$
B_3	Horizontal machine center	Milling face C	Distance	121.3 ± 0.1
B_4	Horizontal machine center	Hinging hole D	Diameter	$\phi 10_{0}^{+0.15}$

With characteristics of B_1, B_2, B_3, B_4 identified as the key product characteristics, interaction of process product quality and system components reliability is established and the associated model of manufacturing system reliability and product quality oriented by product qualified probability is then created. Correspondingly, it is proved to play an important role in the reduction of dimension variations of the cylinder head, the decrease of tooling adjustments, the increase of manufacturing system reliability, and the drop in risk of degradation. Specific modeling processes are shown in Figure 5.

Numerical Example

As shown in Figure 5, reliability analysis and the optimization example for the cylinder head manufacturing systems based on the proposed RQR chain are conducted in the following steps.

Step 1: Create the vector space of key product characteristics as (p_{ix}, p_{iy}, p_{iz}, n_{ix}, n_{iy}, n_{iz}) and ascertain the key feature dimensions of Y_k. To be specific, in order to adapt the automatic detection process, the position vector pi and the orientation vector ni (i = 1, 2, 3, 4) are defined, respectively, to quantify key characteristics of B_1, B_2, $B3$, B_4 facing the three-dimensional space. Table 2 presents the parameters which the vector space (p_{ix}, p_{iy}, p_{iz}, n_{ix}, n_{iy}, n_{iz}) (i = 1, 2, 3, 4) contains.

Based on the theory of SoV [23], transmission of key product dimensions between stations in manufacturing system is simplified as shown in Figure 6. With the major variation u_k and noise of production w_k (k = 1, 2, 3, 4) considered and regardless of noise of measurement γ_k, the produced product dimensions can then be expressed as

$$x(k) = A(k)\,x(k-1) + B(k)\,u_k,$$

$$y(k) = C(k)\,x(k). \tag{22}$$

Parameters of (k), (k), and $C(k)$ in formula (22) are ascertained by variation data of key product dimensions. With reference to what Table 2 has exhibited, use Mathematica software to calculate the four key dimensions as (k) = [13.01, 12.496, 121.18, 10.07].

Table 2: Vector space model of key product characteristics

Number	Key characteristics	n_x	n_y	n_z	p_x	p_y	p_z
1	Hole A	0	1	0	41.5	22.5	0
2	Hole B	0	1	0	14.5.5	22.5	0
3	Face C	0	−1	0	91.5	−15.5	0
4	Hole D	1	0	0	0	0	61.5

Step 2: According to the obtained key dimensions Y_k, determine the product qualified probability $P^c{}_k$ of manufacturing process and the impact $\Pr\{R^Q \mid Z(t)\}$ of process quality Q on reliability of manufacturing system. Assume that the observed original data follow a normal distribution, the related covariance becomes $\sum_{i=1}^{4}\sum_{\substack{j=1\\ j\neq i}}^{4} a_i a_i \mathrm{Cov}(x_i, x_j) = 0$ and μ_f and σ_f^2 are as follows:

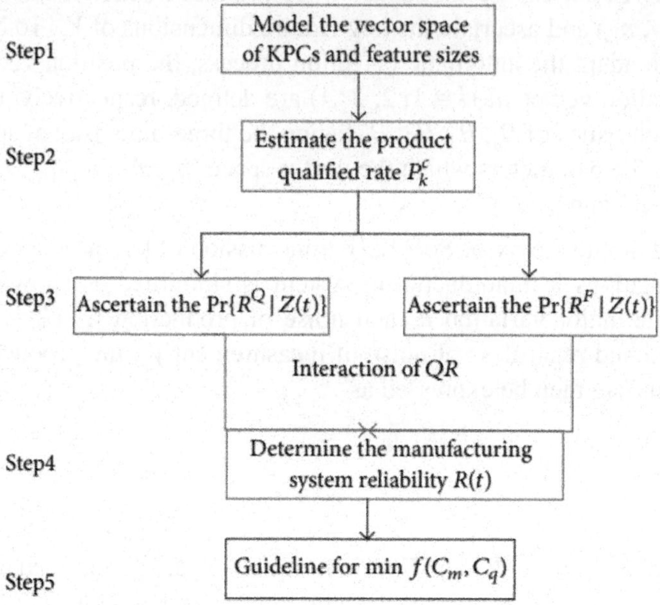

Figure 5: Flowchart of modeling and optimization.

$$\mu_f = f\left(\mu_{x_1}, \mu_{x_2}, \mu_{x_3}, \mu_{x_4}\right) = 0.382,$$

$$\sigma_f^2 = \sum_{i=1}^{4} a_i^2 \sigma_i^2 + \sum_{i=1}^{4} \sum_{\substack{j=1 \\ j \neq i}}^{4} a_i a_j \mathrm{Cov}\left(x_i, x_j\right) = 0.3454. \tag{23}$$

Referring to formula (3), the value of β should be computed as follows:

$$\beta = \frac{\mu_f}{\sigma_f} = \frac{0.382}{\sqrt{0.3454}} = 0.65. \tag{24}$$

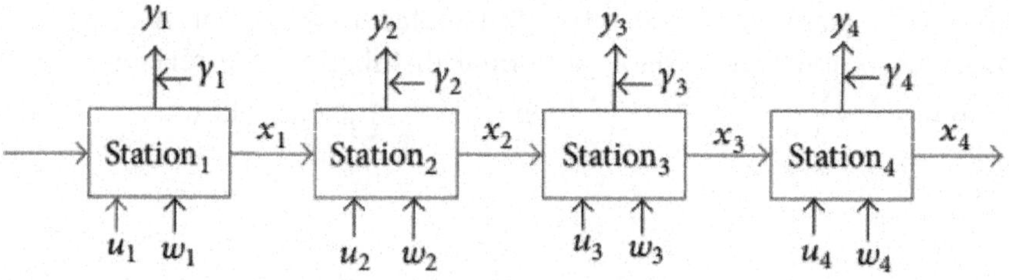

Figure 6: Simplified transmission model of key product dimensions.

And thus the product qualified probability is computed as

$$\Pr\left\{R^{Q} \mid Z(t)\right\} = \Phi(\beta) = \Phi(0.65) = 0.7422. \tag{25}$$

Step 3: Determine the impact of components reliability on manufacturing system reliability $\Pr\{R^F \mid Z(t)\}$ with process quality Q, wear loss, and failure rates of system components quantified, given that the fading rates $w(t_k) = w_0 + \exp(-\varepsilon_k \cdot t_k)$ are parameterized with $w_0 = 5 \times 10^{-5}$ and $\varepsilon_k = 1 \times 10^{-3}$. Meanwhile, the initial failure rate of the system component is $\lambda_{0i} = 6 \times 10^{-6}$ $(i = 1,2,3,4)$ and the interaction coefficients are $\alpha_k = 3 \times 10^{-4}$, $h = 500$, and $\eta = 1.3 \times 10^{-2}$. Set the work time $t$for each station as 200 hours identically. With reference to formula (13), $\Pr\{R^F \mid Z(t)\}$ can be calculated as

$$\Pr\left\{R^{F} \mid Z(t)\right\}$$

$$= \exp\left(\sum_{i=1}^{n}\left(\left[\lambda_0(t) + E\left[\alpha_i (X(i) - m_i)^2\right]\right]\right) \cdot t_i\right)$$

$$\cdot \int_0^{\eta} \exp\left(-h\left(w_0 + e^{-\varepsilon_k \cdot t}\right)\right) d\varepsilon_k = 0.871. \tag{26}$$

Step 4: Based on the product qualified probability $P^c{}_k$ and information of components reliability, the analysis model $(t) = \Pr\{R^F \cap R^Q\}$ for manufacturing system reliability oriented by product inherent reliability is established. Integrate information of process quality $\Pr\{R^Q \mid Z(t)\}$ and components reliability into formula (14), and the reliability of manufacturing system $R(t)$ is finally estimated as

$$R(t) = \Pr\left\{R^{F} \mid Z(t)\right\} \times \Pr\left\{R^{Q} \mid Z(t)\right\}$$

$$= \exp\left(\sum_{i=1}^{n}\left(-\left[\lambda_0(t) + E\left[\alpha_i (X(i) - m_i)^2\right]\right] \cdot t\right)\right)$$

$$\cdot \int_0^{\eta} \exp\left(-h\left(w_0 + e^{-\varepsilon_k \cdot t}\right)\right) d\varepsilon_k \Phi(-\beta)$$

$$= 0.871 \times \Phi\left(\frac{0.382}{\sqrt{0.3454}}\right) = 0.871 \times 0.7422 = 0.646. \tag{27}$$

Step 5: Based on the above results, optimization strategy involved of process quality loss for manufacturing system is analyzed quantitatively.

The intended idea is based on the comparison of the difference between the estimated R_{FQ} from the correlation model, namely, the $\Pr\{R^F \cap R^Q\}$, and the estimated RF irrespective of the correlation between process quality and components reliability, namely, the $\Pr\{R^F \mid Z(t)\}$.

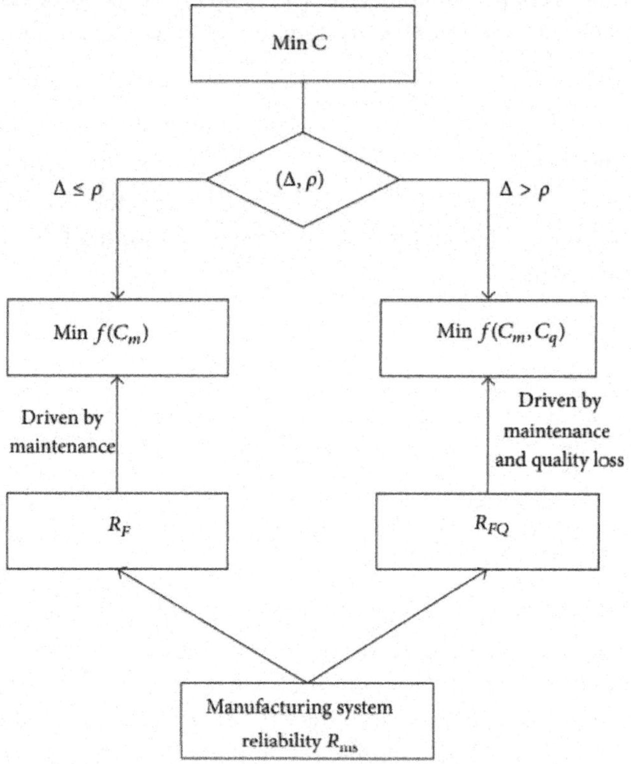

Figure 7: Optimization strategy for the manufacturing system.

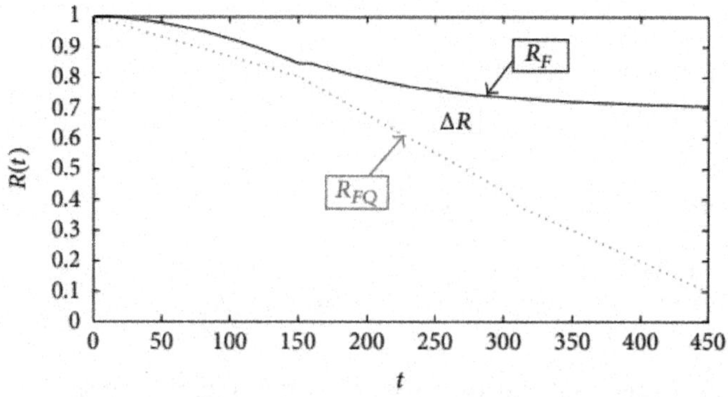

Figure 8: Comparison of manufacturing system reliability for the cylinder head.

With the difference $\Delta = R_F - R_{FQ}$ and the threshold value ρ ascertained, optimization strategy for manufacturing system is shown in Figure 7.

Result Analysis

From the results shown in Section 5.2, it is obvious that when the correlation between process quality and components reliability is not considered, the reliability of manufacturing system is approximately 0.871, whereas it turns to be 0.646 when the correlation is considered. This is to say, manufacturing system reliability is overestimated by 38.9%.

Choose time t as the independent variable; comparison between R_F and R_{FQ} is shown in Figure 8. As shown in Figure 8, the system reliability is often overestimated when we do not consider the interaction between product quality and components degradation or failures. Through the integral operation, the overestimated reliability is 44.97%. Additionally, the estimation error becomes more significant as time increases, which indicates that the cumulative transmission of quality variations does have a profound effect on the system reliability. Furthermore, in order to show the robustness of the proposed approach, sensitivity analysis of the reliability of manufacturing system R on the product qualified probability p_c is simulated, and the result is shown in Figure 9.

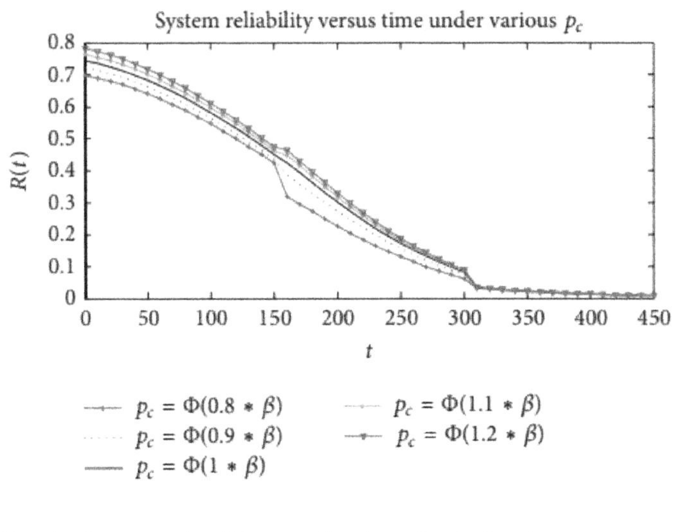

Figure 9: Sensitivity analysis of R on p_c for the cylinder head.

As shown in Figure 9, five levels of pc have been set to analyze the significant impact of the cylinder head qualified probability on the reliability of the manufacturing system (t). With a series of ±20% shift from the basic $\Pr\{R^Q| \quad Z(t)\} = \Phi(\beta) = \Phi(0.65) = 0.7422$, namely, $p_c = \Phi(0.8 * \beta)$, $p_c = \Phi(0.9 * \beta)$, $p_c = \Phi(1 * \beta)$, $p_c = \Phi(1.1 * \beta)$, and $p_c = \Phi(1.2 * \beta)$, five reliability curves of the manufacturing system are plotted versus time t under different levels of the product qualified probability pc to demonstrate the correlated sensitivity and robustness. And the synthetical simulation result shows that a high level of the proposed cylinder head qualified probability can usually guarantee a high level of the system reliability. Accordingly, the model of manufacturing system reliability based on the product qualified probability would help the reliability engineer to form a more objective and accurate mathematical model to analyze comprehensively the reliability and performance for manufacturing system. Moreover, the appropriate optimization strategies covering the maintenance for system components and adjustment of process scheme can be got, which should provide clear goals and directions to continuously improve the manufacturing system reliability.

CONCLUSION

In this paper, the RQR chain which could fuse the quality data and reliability data in manufacturing system reliability optimization is put forward, and the product qualified probability is introduced to quantify the impacts of quality variation in manufacturing process on the reliability of manufacturing system at the first time. Furthermore, a novel mathematical model of manufacturing system reliability is established based on the product qualified probability and RQR chain. Finally, the reliability and maintenance optimization strategy for manufacturing system is analyzed in view of total cost of maintenance costs and quality loss. The application result demonstrated that manufacturing system reliability tends to be overestimated if the mentioned interaction between product quality and manufacturing system reliability is omitted. The overrated value of the reliability of manufacturing system will not only deteriorate the produced product reliability and may also lead to a wrong maintenance strategy or miss the best opportunity for system maintenance.

To conclude, it is critically essential to consider the coeffects between process quality and system components reliability when modeling and assessing the reliability of manufacturing system. For future research, the following topics should be further expounded.

- The improvement of the reliability optimization model based on the quality loss for different type of manufacturing system is needed. The coefficients of the optimization model are different for different manufacturing system; therefore, how to estimate accurately the coefficients from the big data from manufacturing system design, operation, and maintenance is planned.

- The quantitative mathematical relationship of the manufacturing system reliability, manufacturing process quality, and the produced product reliability should be established successively. Specifically, the mathematical impact of the manufacturing system reliability on the produced product reliability should be constructed clearly, which should provide a solid foundation for the integrated reliability and maintenance optimization framework of the various types of manufacturing system.
- In the last perspective research, we consider the aspect of reliability modeling and assessment in the design and setup of manufacturing system. The reliability level is determined in the design process of the manufacturing system, in order to satisfy the ever-increasing stringent quality and reliability requirements, the reliability design should be integrated with the functional design of manufacturing system, and new design theory like Axiomatic Design should be adopted into reliability design of manufacturing system.

ACKNOWLEDGMENT

This research was supported by Grant 61473017 from the National Natural Science Foundation of China.

REFERENCES

1. Y. He, W. Linbo, Z. He, and M. Xie, "A fuzzy TOPSIS and rough set based approach for mechanism analysis of product infant failure," Engineering Applications of Artificial Intelligence, 2015.
2. M. Colledani, T. Tolio, A. Fischer et al., "Design and management of manufacturing systems for production quality," CIRP Annals: Manufacturing Technology, vol. 63, no. 2, pp. 773–796, 2014.
3. P. Murthy, "New research in reliability, warranty and maintenance," in Proceedings of the 4th Asia-Pacific International Symposium on Advanced Reliability and Maintenance Modeling (APARM '10), pp. 504–515, McGraw-Hill International Enterprises, Wellington, New Zealand, December 2010.
4. R. Jiang and D. N. P. Murthy, "Impact of quality variations on product reliability," Reliability Engineering & System Safety, vol. 94, no. 2, pp. 490–496, 2009.
5. J. Li and N. Huang, "Quality evaluation in flexible manufacturing systems: a Markovian approach,"Mathematical Problems in Engineering, vol. 2007, Article ID 57128, 24 pages, 2007.

6. J. Li and J. Lei, "Integration of manufacturing system design and quality management," IIE Transactions, vol. 45, no. 6, pp. 555–556, 2013.
7. R. R. Inman, D. E. Blumenfeld, N. Huang, J. Li, and J. Li, "Survey of recent advances on the interface between production system design and quality," IIE Transactions, vol. 45, no. 6, pp. 557–574, 2013.
8. N. Li and J. Ni, "Reliability estimation based on operational data of manufacturing systems," Quality and Reliability Engineering International, vol. 24, no. 7, pp. 843–854, 2008.
9. Y.-K. Lin and P.-C. Chang, "System reliability of a manufacturing network with reworking action and different failure rates," International Journal of Production Research, vol. 50, no. 23, pp. 6930–6944, 2012.
10. G.-D. Li, S. Masuda, D. Yamaguchi, and M. Nagai, "A new reliability prediction model in manufacturing systems," IEEE Transactions on Reliability, vol. 59, no. 1, pp. 170–177, 2010.
11. Y. Chen and J. Jin, "Quality-reliability chain modeling for system-reliability analysis of complex manufacturing process," IEEE Transactions on Reliability, vol. 54, no. 3, pp. 475–488, 2005.
12. Y. Chen and J. Jin, "Quality-oriented-maintenance for multiple interactive tooling components in discrete manufacturing processes," IEEE Transactions on Reliability, vol. 55, no. 1, pp. 123–134, 2006.
13. F. Zhang, J. Lu, Y. Yan, S. Tang, and C. Meng, "Dimensional quality oriented reliability modeling for complex manufacturing process," International Journal of Computational Intelligence Systems, vol. 4, no. 6, pp. 1262–1268, 2011.
14. K. Rafiee, Q. Feng, and D. W. Coit, "Reliability modeling for dependent competing failure processes with changing degradation rate," IIE Transactions, vol. 46, no. 5, pp. 483–496, 2014.
15. N. Li, F. T. S. Chan, S. H. Chung, and A. H. Tai, "An EPQ model for deteriorating production system and items with rework," Mathematical Problems in Engineering, vol. 2015, Article ID 957970, 10 pages, 2015.
16. X. Gong, Y. Feng, H. Zheng, and J. Tan, "An adaptive maintenance model oriented to process environment of the manufacturing systems," Mathematical Problems in Engineering, vol. 2014, Article ID 537452, 10 pages, 2014.
17. L. Tlili, M. Radhoui, and A. Chelbi, "Condition-based maintenance strategy for production systems generating environmental damage," Mathematical Problems in Engineering, vol. 2015, Article ID 494162, 12 pages, 2015.

18. Z. Hajej, N. Rezg, and A. Gharbi, "Forecasting and maintenance problem under subcontracting constraint with transportation delay," International Journal of Production Research, vol. 52, no. 22, pp. 6695–6716, 2014.
19. H. Zied, D. Sofiene, and R. Nidhal, "Joint optimisation of maintenance and production policies with subcontracting and product returns," Journal of Intelligent Manufacturing, vol. 25, no. 3, pp. 589–602, 2014.
20. Z. Hajej, N. Rezg, and A. Gharbi, "A decision optimization model for leased manufacturing equipment with warranty under forecasting production/ maintenance problem," Mathematical Problems in Engineering, vol. 2015, Article ID 274530, 14 pages, 2015.
21. Z. Hajej, S. Turki, and N. Rezg, "Modelling and analysis for sequentially optimising production,maintenance and delivery activities taking into account product returns," International Journal of Production Research, vol. 53, no. 15, pp. 4694–4719, 2015.
22. L. Mifdal, Z. Hajej, and S. Dellagi, "Joint optimization approach of maintenance and production planning for a multiple-product manufacturing system," Mathematical Problems in Engineering, vol. 2015, Article ID 769723, 17 pages, 2015.
23. J. Shi, Stream of Variation Modeling and Analysis for Multistage Manufacturing Processes, CRC Press, Taylor & Francis, 2007.

Chapter 3

WORKERS AND MACHINE PERFORMANCE MODELING IN MANUFACTURING SYSTEM USING ARENA SIMULATION

Kassu Jilcha, Esheitie Berhan and Hannan Sherif

Addis Ababa University, Addis Ababa Institute of Technology, Addis Ababa, Ethiopia

INTRODUCTION

Industry Performance improvements arising from increased manufacturing integration continues to be one of the primary competitive issues in current days. Faced with ever-increasing challenges such as the globalization, increased world competition, and increased customer expectations, companies are pursuing strategies to improve their performance and reduce their costs. Discrete-event modeling and simulation (DES) is a popular tool in widely varying fields for identifying and answering questions about the effects of changes on processes. Recent research into manufacturing systems integration has identified the need for effective positioning of business objectives down through the organization and the subsequent measurement of performance in critical areas as key elements of sustainable competitive advantage [1-7]. One of the best tools that can indicate the performance of any company is simulation software. The simulation software can be divided in to bigger group Simulation Language and Simulators. The simulation language is programming based using simulation software. The simulator allows one to build a model of the desired system by using before- made modeling constructs [8]. In Simulation, language software the model is developed creating a program syntactically or graphically using language's modeling constructs. These types of simulation software are very flexible tools but the user needs to know programming concepts and longer modeling times [8,9]. Thus, simulation can be either discrete event or continuous simulation. In continuous model, the state of the system can change continuously over time. In discrete model, though changes can occur only at separated points in time [10].

During the past decade, the manufacturing industry has undergone a dynamic transformation. Recently, traditional manufacturers are inherently subject to high-mix, low volume manufacturing as a business model [11]. In this paper, discrete event simulation is used to model workers and machine performance in Ethiopia Plastics Industry. Highly industries rely on workers and machine performance since productivity depends on workers and machine availability. A company will not be able to participate and will not survive in a long-term perspective if employees do not attribute a high significance to such organization goals. However, poor workers and machine and machine performance can be recognized some of the factors are of high load of work, financial stress, and illness. Therefore, workers and machine performance, as this function is essential required for huge value to the organization in maintaining and strengthening its business and revenue growth [12-16].

¨Multi-skilling of workers can be achieved by cross-training. Crosstraining is a process in which workers are trained on the tasks, duties, and responsibilities of multiple tasks in a specific work cell or work area [14]. Worker assignment and involvement within a labor-intensive cell is a major role in the performance of the cell [17]. The first model assigns multi-skilled workers to existing cells while the second model finds the optimal operator assignment within a cell [15]. The performance of the workers can be determined from the arena simulation tools to take an action in line with the other techniques and tools presented in the previous researches.

The problem observed in this case company is the lack of experts and variation on processing the PVC pipe in adjusting the heat treatment and low capacity utilization of the workshop which causes variation to the specified product. During this, the number of recycled products is higher per days. The industry can fail to achieve the goal for efficient and economical production. Capacity includes all resources such as machineries, workers etc. of the factory.

Therefore, the objective of this paper is to provide industry with a complete set of tools, techniques and procedures to allow examining of existing workers and machine performance systems. These techniques helps in applying simulation system model which could be used as before an existing system is altered or a new system built, to reduce the chances of failure to meet specifications, to eliminate unforeseen bottlenecks, to prevent under or over-utilization of resources, and to optimize system performance.

MATERIALS AND METHODS

The study was conducted by considering different materials and methods to achieve the goal of this study. The literature reviews and arena software applications were utilized to come up with good solution. Simulation, in particular discrete event simulation, has still not gain industries wide acceptance as decision support tool for performance evaluation. The researchers have provided the example of the application of discrete event simulation in evaluating performance of a production line in the case company. The data was collected from an Ethiopian Plastic industry production line and the simulation model is built using arena software simulation. A simulation model provides a visual animation of the line to see how product will flow through the line. An ARENA® simulation model was developed, verified, and validated to determine the daily production and potential problem areas for the various production levels in the plastic industry.

DISCUSSION AND RESULTS

Production Process Flow

In general, there are 5 main workshops in Ethiopia plastic industry. Raw materials in the form of garden hose and master batch are supplied by foreign countries like Asia and in domestic like Elsewady Cables Ethiopia and directed to raw material store for storing and quality checking.

They will be sent to two operating process continuous and injection section and turn on heat treatment and take the melted plastic with screw thread then forced into mold. The bundle will be prepared by hauloof. After the product is finished it will be inspected and then packaging will be done. Finally; the product will be kept in finishing storage as shown in Figure 1. The other process is continues (extrusion) process flow which begins at raw material receiving, and balling unit machine as indicated by Figure 2.

Data Analysis

The collected data through the means of interviews, direct observation and company document review were analyzed. Fitting Input Distribution through the input analyzer is used to identify fitted statistical distribution [18]. It is used to evaluate the distribution's parameter and calculates a number of measures of the data. To select which type of distribution to use, the authors have compared the square error of each distribution. The larger the square error value, the further away the fitted distribution is from the actual data. The data was analyzed by using real recorded time from the operation areas of the selected product in the case company section. These data supported the analysis of validation, verification and number of replication estimation as discussed in the next sections [7].

Model Assumption

In Ethiopia Plastics Industry, there are three shifts per day. Workers distribution through the specified time per shifts can be delayed by different kind of incidents. The actual working time for whole system is 8:00 working hours per day. Since it is not expected that a person will work without some interruptions, the operators may take time for their personal needs. Assumptions are taken for modeling the variation of workers performance at time to rest, machine failure, and power off. The model for one shifts are modeled and analyzed. In the study only polyvinyl chloride (PVC) pipe line is selected. Raw material is assumed as a constant input, no interruption of operation is occurred in continuous and random for injection. The manufacturer working time is 8:00 working hours per day. Distribution is fitted to input analyzer; it evaluates the distribution's parameter and calculates a number of measures of the data. In order to select which type of distribution is used, the author has compared the square error of each distribution. The larger the square error value, the further away the fitted distribution is from the actual data (19). Therefore the following fit all summary orders in worker performance distribution from smallest to largest square error. From the table it can be seen that triangular is the one with smallest square error and thus it is selected. Line indicates that the data normal distribution and the blocks are the actual data's situation histogram expression when compared to the normal distribution. The normal distribution is skewed to the right (Tables 1-3, Figure 3 [18,19]).

Model formulation

In modeling design the type of analysis is performed on workers and machine performance. The products that are selected are the one those most often produced. They are considered as one of the main products of Ethiopia Plastic Industry in this case company. The working period in Ethiopia Plastic Industry is five days and a half day within eight working hour consideration. This means the working days from Monday to Friday 8:00 working hours and on Saturday 4:00 working hours are utilized in the company. The three models for the arena simulations are workers distribution models, continuous extrusion models, and injection models as shown in Figures 4-6.

Model verification

Verification assesses the correctness of the formal representation of the intended model, by inspecting computer code and test runs, and performing consistency checks on their statistics [20]. More specifically, verification consists in the main of the following activities:

Inspecting simulation program logic

- Performing simulation test runs and inspecting sample path trajectories. In particular, in a visual simulation environment (like Arena), the analyst inspects both code printouts as well as graphics to verify (as best one can) that the underlying program logic is correct.
- Performing simple consistency checks, including sanity checks as well as more sophisticated checks of theoretical relations hips among predicted statistics.
- In simulation modeling, each part of model run with different set of inputs and the obtained outputs were compared by using throughputs and little's formula with real outputs.

Model validation

Validation activities are critical to the construction of credible models. The standard approach to model validation is to collect data (parameter values, performance metrics, etc.) from the system under study, and compare them to their model counter parts. These parameters will prove the validity of the data and simulation.

Number of replication estimation

A good design of simulation replications allows the analyst to obtain the most statistical information from simulation runs for the least computational cost. In particular, minimizing the number of replications and their length is necessary to obtain reliable statistics. In order to decide the number of replication the model must run some initial set of replication so that sample average, standard deviation and confidence interval are computed. This initial set of replication is shown in Table 4.

For PVC pipe continuous extrusion production line initial replication (no) is 6 and the initial half width (ho) is 21.40. Assuming that the h= 0.25, taking 95% Confidence Level.

DF= n-1= 6-1= 5

$$n \cong t^2 n-1, 1-a/2 \frac{s^2}{h^2}$$

$$n \cong t^2 6-1, 1-0.05/2 \frac{3.97^2}{0.25^2}$$

$$n \cong t^2 0.025, 5 \frac{3.97^2}{0.22^2} \cong (2571)2 * \frac{3.97^2}{0.25^2} = 1,666$$

By considering this equation the number of replications is 1,666

CONCLUSION

In this study worker and machine performance in manufacturing system using simulation has been used. The simulation is done by using Arena software. The performance measure used is throughput and it is directly related to waiting time; work in process, resource or capacity utilization. Simulation models were built for the continuous (extrusion) and injection workshops separately and finally a model is prepared for the purpose of conducting the computer based simulation. The problems identified are the throughput time from the existing system is low because of occurrence of bottlenecks, and waiting time identified. Workers agent should be assigned on the machinery and workers performance to focus on each operating system due that the number of recycle products can be reduced. Otherwise, the company continues like this for a long time and it will be difficult for them to compete in today's market. Therefore, to improve the existing performance of the company, resource utilization improvement is a mandatory.

REFERENCES

1. Suer GA (1996) Optimal Operator Assignment and Cell Loading in Labor-Intensive Manufacturing Cells. Computers & Industrial Engineering 31: 155-158.
2. Hasgul S (2005) Winter Simulation Conference.
3. Marco Semini, HakonFauske (2006) Applications of Discrete-Event Simulation to Support Manufacturing Logistics Decision-Making: A Survey. Proceedings of the 2006 Winter Simulation Conference.
4. JuhanniHeilala (1999) Use of Simulation in Manufacturing and Logistics systems planning. VTT Industrial Horizon, Espoo.
5. David WK, Randall S, Nancy Z (2010) Simulation with Arena (5thedn) McGraw-Hill.
6. Macintosh CA (1992) Research in Manufacturing Systems Integration, Integration in Production Management Systems. Pels,Worthman (eds.) Elsevier.
7. Paul B, Bennett LF, Linus ES (1986)A Guide to Simulation.2nd Edition, Springer-Verlag New York Berlin Heidelberg London Paris Tokyo.
8. Francisco EM (2002) Application of SIMAN ARENA Discrete Event Simulation Tool in the Operational Planning of a Rail System. University of Puerto Rico Mayagüez Campus.

9. William PB, Henrich RG, Douglas YP (1994) An Evolutionary Model of Organizational Performance. Strategic Management Journal 15: 11-28.
10. Tayfur A, Benjamin M (2007) Simulation Modeling and Analysis with Arena. Rutgers University, Piscataway, New Jersey, Elsevier Inc.
11. Ahad SK, Hamid S (2006) Simulation Intelligence and Modeling for Manufacturing Uncertainties. Proceedings of the 2006 Winter Simulation Conference.
12. Ravindran APD (1987) Operations Research: Principles and Practices. John Wiley & Sons.New York.
13. Umit SB (1995) Modelling of performance measurement systems in manufacturing enterprises. International Journal of Production Economics 42: 137-147.
14. Theodore TA (2011) Introduction to Discrete Event Simulation and Agent-based Modeling: Voting Systems, Health Care, Military and Manufacturing.
15. Karaman AS (2007)Performance Analysis and Design of Batch Ordering Policies in Supply Chains.
16. McDonald T, Eileen VA, Kimberly E (2011) Utilizing Simulation to Evaluate Production Line Performance Under Varying Demand Conditions. Proceedings of the 2011 IAJC-ASEE International Conference
17. McDonald TN (2004) Analysis of Worker Assignment Policies on Production Line Performance Utilizing a Multi-skilled Workforce. Dissertation for Virginia Polytechnic Institute and State University for the degree of Doctor of Philosophy in Industrial and Systems Engineering, Blacksburg, Virginia.
18. Arturo M, Veruzcka M (2003) Application of enterprise models and simulation tools for the evaluation of the impact of best manufacturing practices implementation. Annual Reviews in Control27: 221-228.
19. Phillips DT, James JS (1987) Operations Research: Principles and Practice (2ndedn.) John Wiley & Sons (2014) Statistical distribution. ARENA Basic Edition User's Guide

Chapter 4

USABILITY EXPERIMENTS TO EVALUATE UML/ SYSML-BASED MODEL DRIVEN SOFTWARE ENGINEERING NOTATIONS FOR LOGIC CONTROL IN MANUFACTURING AUTOMATION

Birgit Vogel-Heuser

Institute of Automation and Information Systems, Technische Universität München, Munich, Germany

ABSTRACT

Many industrial companies and researchers are looking for more efficient model driven engineering approaches (MDE) in software engineering of manufacturing automation systems (MS) especially for logic control programming, but are uncertain about the applicability and effort needed to implement those approaches in comparison to classical Programmable Logic Controller (PLC) programming with IEC 61131-3. The paper summarizes results of usability experiments evaluating UML and SysML as software engineering notations for a MDE applied in the domain of manufacturing systems. Modeling MS needs to cover the domain specific characteristics, i.e. hybrid process, real time requirements and communication requirements. In addition the paper presents factors, constraint and practical experience for the development of further usability experiments. The paper gives examples of notational expressiveness and weaknesses of UML and SysML. The appendix delivers detailed master models, representing the correct best suited model, and evaluation schemes of the experiment, which is helpful if setting up own empirical experiments.

INTRODUCTION

Today, manufacturing automation systems (MS) mostly consist of PLC-based control systems [1] programmed in the languages of the IEC 61131-3 standard [2] . Since the proportion of system functionality that is realized by software is increasing [3] , concepts for supporting automation engineers in handling software complexity are strongly required. Furthermore, spatially distributed MSs tend to result in a spatial distribution of software [4] -[6] , i.e. networked automation systems (NAS). One key challenge of manufacturing automation companies in high wage countries is to increase efficiency, effectiveness and quality in design of software engineering for MS to shorten engineering and start-up time and ease maintenance. To reach this goal, model driven software engineering (MDE) including code generation is a promising systematic approach [1] . A variety of different modeling notations (general and domain specific ones) were developed and defined by academia and/or tool sup- pliers during the last decades to improve efficiency and effectiveness and to increase software quality in manufacturing automation, e.g. Vyatkin [7] -[9] , Fantuzzi, et al. [10] -[12] , Thramboulidis, et al. [3] [13] [14] , Vogel- Heuser, et al. [15] [16] or Estévez, et al. [17] [18] .

Many industrial companies and researchers are looking for more efficient MDE in software engineering of MS, but are uncertain about the applicability and effort including training and reengineering of component libraries and workflow needed to implement those approaches in their companies. In the last decade many new approaches, e.g. notations as UML and SysML and tools to support these approaches were developed and presented to industry, to improve software quality and efficiency as well as maintainability and evolution including more or less the typical requirements as real time, communication and partially hybrid process characteristics or networked automation systems. The challenge for manufacturing automation companies is to find the most appropriate MDE approach fitting to the companies' particular requirements, e.g. market requirements, customer relation (what to get from the customer and what to deliver), workflow requirements, qualification of personnel, budget and so forth. Consultants try to bridge this gap and support automation companies specifying their requirements and evaluating available approaches and/or tools. In a next step a beta-test may be conducted which is always short in time and delayed by higher prioritized "real" projects. Often poor usability of tools, missing features or missing modular and flexible training units lead to a rejection of a MDE approach [19] , which would be applicable and beneficial if systematically selected and introduced. But the risk to spend too much time of experienced application engineers for evaluating a probably inappropriate notation and/or tool is often estimated as too high to allow a systematic and exhaustive evaluation.

To overcome this drawback and enable selecting the best notation which fits well to a company's needs, this paper provides a discussion of different ways for usability evaluation in Section 2 including usability evaluation methods and procedures that are well-established and have successfully been applied for decades, e.g. in the automotive domain [20] and in nuclear safety research [21]

. In the following Section 3, aspects and rules to be considered when designing experiments are investigated in detail. These proposed aspects are exemplified by a variety of experiments conducted in the last decade on usability of software engineering in MS presented in (Section 4). The paper closes with a summary and discussion in Section 5 and 6, respectively. The appendix gives three examples of the evaluation of subjects' models. Appendix A shows subjects' models from the first experiments highlighting the difficulties in understanding class as a structural mechanism to foster modularity and reuse using UML 1.4. Appendix B shows an example of subject's UML model (Evaluation scheme E1) compared to a master model also highlighting the calculation of the later used complexity measure WMC. Appendix C explains the different abstraction mechanism between subjects' solution and master model.

METHODS IN USABILITY EVALUATION

Up to now the standard procedure in industry when testing the applicability of different modeling notations in automation technology is a consultation of experts (application engineers) working in the industrial sector of interest [22] . Alternatively—or as a supplement to that practice—end users are questioned, e.g. the programmers of PLCs [23] having in fact the same drawbacks. From a methodological point of view, the consultation of experts can either be individually (i.e. interviews) or in an interactive group setting (i.e. focus group or case studies in workshops). Both approaches measure a subjective assessment of the respondents, i.e. their attitudes, opinions and knowledge can be gained—not their objective behavior. Focus groups as well as case studies operate with such small numbers of experts to start exploring usability issues. Focus groups require a moderator asking questions and moderating the discussion. In case studies an observer is required monitoring and observing the ex- perts actions. Gained insights can serve as an inspiration for further, more detailed, in-depth studies with more explicit hypotheses. Consequently, the consultation of experts or end users is not the only appropriate way to investigate the applicability of a notation but to identify which notations might be relevant and/or to gain first qualitative results on weaknesses and drawbacks. A major advantage of focus groups compared to individual interviews is a more natural atmosphere that ideally leads to increased talkativeness and openness of the participants [24] . An example of a focus group evaluation is given in [25] . In contrast, individual interviews are easier to perform (especially with busy industrial experts), as the persons can be questioned at different times and different places, i.e. no common appointment has to be found. Mostly, a very limited number of experts is available for evaluation, and consequently, a representative quantitative opinion cannot be gathered. However, an evaluation by experts, as for example described in [26] , can act as a beta-test of a prototypical tool indicating the opinion trend. The same challenge occurs with case studies testing industrial experts and/or researchers as e.g. de- scribed in [27] and [28] . In both studies, a significant small group of participants were observed during courses on programming in accordance to the IEC 61499 compared to IEC 61131-3 standard and an Object Oriented approach. Conclusions based on observed aspects can only serve as indicators. Consequently, experiments with a higher number of subjects are required to gain objective, detailed and quantitative results for specific research questions. The in depth investigation of research questions like the applicability of novel modeling notations or programming paradigms presuppose that test persons get familiar with the novel notation or paradigm. The introduction of the required knowledge is very time-consuming and consequently not applicable for a large number of experts, i.e. minimum 15 - 20 per compared notation, required for significant, objective results. Further- more, the comparability of results of different experts with different previous knowledge is very difficult to handle. Thus, such research questions (e.g. the usefulness of a certain new notation) can only be quantified with laboratory experiments, where comparable testing conditions can be created and where, due to the sheer number of tested subjects (plus respective grouping), their individual characteristics and preferences can be considered. Such relevant testing conditions and constraints which have to be considered during experimental design are discussed in detail in the following section. Furthermore, their relationship to usability and how to quantify experimental results in order to deduce quantifiable usability results is presented.

FACTORS FOR EXPERIMENTAL DESIGN IN MS

Patig [29] [30] and Gemino and Wand [31] present results of various experiments on the usability of modeling notations in requirement engineering and classical software engineering introducing relevant factors for successful experimental design. Following an experimental approach the experimental design needs to be fixed ad first, choosing the constraints of the experiment, the so called affecting variables (Figure 1), i.e. the process to be controlled, the engineering task as well as the duration of the whole experiment depending on the temporary availability of the subjects and the subjects themselves. These affecting variables are described in Section 3.1. Hereafter we discuss the affected variables or usability requirements for experimental design in Section 3.2, showing the measures to evaluate the different notational approaches. Thereby, we draw both on [29] [30] and [31] as well as our earlier reflections [32] . On that basis, we adapted and enlarged the extent of variables with focus on the specific requirements in design and maintenance of MS.

Overview of Affecting Variables

According to the best of the authors' knowledge, the main three influencing factors seem to be (1) the task, (2) the preceding training and (3) the tested subjects (Figure 1). Because subjects' attendance for training and expe- riment is a prerequisite, duration is not only a training constraint but also a constraint for the whole experiment including the performances of the engineering task. For experiments with students or apprentices two days is most promising from an organizational point of view.

Task

To describe the requirements from the technical process point of view the automation task is introduced and separated from the engineering task performed by an application engineer using notations and tools.

1) Automation Task

When designing an experiment, e.g. in order to evaluate a given notation for an enterprise, the task to be automated, i.e. the automation task, and its characteristics has to be clarified. The characteristics should be selected similar to characteristics of the real automation tasks in the company's daily business. Due to the complexity of real industrial automation tasks, the main challenge in the experimental set-up is to develop an appropriate automation task with adjusted complexity the subjects can cope within the limited time frame of the experiment and the required training.

As measure for complexity of an automation task, the number of inputs and outputs (I/O) is a familiar metric in cost estimation during preparation of an offer distinguishing between analog (closed loop) and digital (logic design) inputs and estimating the complexity of the closed loop control with a factor (see column "automation task"). However, it is far too simple to reduce the complexity of an automation tasks to the number of I/Os be- cause the automation task itself involves even more complexity. To derive the complexity of the automation task, it is necessary to describe the automation task itself with a notation. Therefore, then notational influence may lead to different complexity measure for the same control task. In the following measures for control task com- plexity correlated to the chosen notation are introduced.

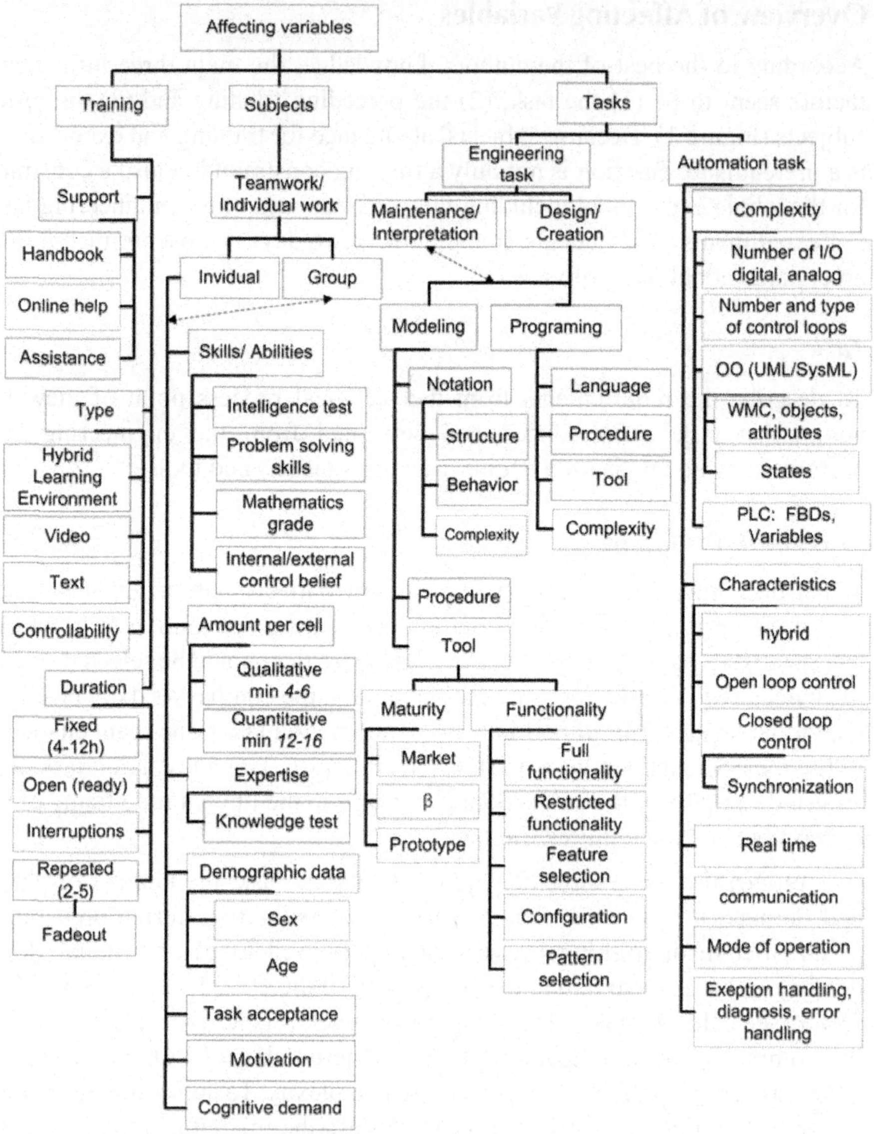

Figure 1. Affecting variables (dashed lines show dependencies between group and substructure and maintenance/interpretation and substructure).

Lukas et al. [33] introduce different complexity measure, i.e. the size (number of operation and number of state variables), modularity and interconnectedness and applied those in logic control. Frey and Litz [34] introduced complexity metrics for Petrinet using besides others an adapted McCabe metric. Venkatesh et al. [35] proposed to count the number of elements required to represent a certain program in order to measure its complexity as well as Lee and Hsu [36] , who converted the programs in question into Boolean expressions by using if-then transformations and, afterwards, rated the programs' complexity by comparing the calculated values.

For applying an object-oriented notion for describing the automation task, Chidamber and Kemerer developed a set of metrics of OO design [37] , e.g. weighted methods per class (WMC) which is a measure for class com- plexity used in this paper (see column control complexity in Table 1): in order to calculate the WMC of a pro- gram, the cyclomatic complexity measure of each method is summed up for all classes, cf. [38] . When applying classical IEC 61131-3 code to describe an automation task, the number of FBDs and their instances in case of IEC 61131-3 are a familiar measure [39] .

Besides those metrics which rely on the description of the automation task for deriving its complexity, also its characteristics like the type of control loop, i.e. logic control, closed loop control (with synchronization) or a technical process requiring both, called hybrid in the following can be taken into account. Furthermore require- ments on automation systems, which have to be fulfilled are introduced: real time requirements, communication requirements between different controllers, and networked automation systems (NAS) as a class of systems with real time and communications requirements because the code and functionality is distributed onto different au- tomation devices.

Besides the regular machine control function, diagnosis, exception handling, visualization and other functio- nality need to be developed. "In fact, industry folklore suggests that approximately 90% of the overall control logic is used for exception handling" [40] . During the last years, the authors' team analyzed real PLC code from several MS companies and realized that 6% - 10% of the lines of code are dealing with diagnosis and safety [41] . As a consequence, the mode of operation (EN 13128 [42] : auto, hand, manual etc.) as well as diagnosis includ- ing error handling (according to [38] and [41]) are typical automation tasks which provide another possible clas- sification.

2) Engineering Task

The category engineering task describes the task to be solved by the human in the experiment, e.g. model (UML, SysML) or program (IEC Code) to control the automation task. Gemino und Wand [31] distinguish two types of human tasks in automation and control:

- design and creation versus
- understanding and analysis (being typical for maintenance tasks in MS).

For both types of tasks, it must be between modeling and programming (dashed lines in Figure 1). Moreover, maturity and functionality of tool support have to be considered as influencing factors. As tool classification three maturity levels and four different functionality types are proposed. Because the comparison of different notations in MDE is focused in this paper the modeling notation as such and its measures should be discussed in more detail.

The notation needs to fulfill the requirements given by the automation task, the life cycle model and the engineering task allowing modeling structure and behavior. Different modeling notations also possess different complexities. Calculation schemes are proposed by e.g. Recker et al. 2009 [42] and Rossi and Brinkkemper [43]. The expressiveness of this measure is limited to the complexity of the pure notation calculated on the number of its elements.

$$C(M) = \sqrt{O^2 + R^2 + P^2} \tag{1}$$

With O being the number of object types, R describing the relationship types and P the property types of a method, C(M) is the resulting complexity of a heterogeneous modeling language.

Schalles [44] compared UML activity diagrams for behavioral modeling and UML class diagrams for struc- tural modeling, taking into account the high complexity differences (Table 2).

To allow the comparison of notations' complexity, the complexity of the later discussed notations is calculated, too. The complexity of the UML class diagram (Table 2) is nearly double compared to IEC 61131-3, SysML- AT (see also [45]) and CFC showing that between the typical MS notations the difference is very small.

Subjects

Subjects' qualification and experience is essential for the outcome of the experiment. Sierla et al. [28] and Ha- jarnarvis et al. [46] included industrial experts. Sierla introduced teamwork with clearly separated tasks similar to a real project team in industry.

Many papers on programmers' competencies in modeling and informatics systems application are related to competence models, e.g. [47] -[49] . Usually, competencies in this context are understood as abilities, skills, and knowledge—a perspective which is still prominent in most Anglo-American research on competencies. An ex- ample is provided by Curtis [50] , who proposed that programming results depend on individual personal factors and mental abilities. His model covers intellectual aptitudes, the knowledge base, cognitive styles, the motiva- tional structure, personality characteristics, and behavioral characteristics. Although Curtis did not empirically test his model, the factors show at least facial validity.

Other approaches for gaining insights into competencies required for different programming approaches or skills analyze interviews from experts in the questioned domain [51] or evaluate programmers' behavior when performing certain tasks, e.g. programming tasks [52] or debugging tasks [49].

Table 1: Overview of the experiments

						Criteria				
	Compared MDE Approaches/ Notations	Engineering Support	Tool Support	Control Task	Training Type (Duration h)	Subjects' Qualification; Min. Number per Cell	Design Overall Evaluation	Results	Results Subjective Questionnaire	Results for Further Notation/Tool Development
O.1	IEC 61131 LL, Petrinet, SIPN, mFSM Lucas and Tilbury [67], Lucas [68]	N	LL (from manufacturer of test bed) mFSM (paper and pencil)	Flexible manufacturing test bed Logic Control 30DI/O (n.A.)	n.A.	LL professional, PN, SIPN, mFSM by students		Estimation of development times based on low level user operations, and observation of engineers performing their task LL-405 min PN-1110 min Mfsm-1500 min	n.A.	Schema for estimation of time based on low level user operations
O.2	IEC 61131 LL, Petrinet, SIPN, mFSM Lucas and Tilbury [33]	N	LL (from manufacturer of test bed) mFSM (paper and pencil)	Flexible manufacturing test bed Logic Control, DI/O 30 4 scenarios (n.A.)	n.A.	LL professional, PN, SIPN, mFSM by students		Number of operations, state variable, modularity, interconnectness	n.A.	LL small but very interconnected, mFSM most modular, although largest
O.3	IEC 61499 vs IEC 61131-3 plus design pattern repository Strömman *et al.* [27]	N, P	Market tool, repository	Lifter unit Logic Control (-)	Presentation of design pattern, repository and FBDs	20 Professional automation designers and researchers		Models analyzed, and written feedback 3	Tape recorded interviews in companies as preparation each 60 min.	Event driven model experience relevant, guidelines for reusable software, environment that fosters collaboration and exchange of information
O.4	IEC 61499, IEC 61131, OO Sierla *et al.* [28]	N,P	Market tool, adapted 61499 framework, repository	(hyb) Per notation 1 week course for training incl. experiment All approaches different sequence	Set of domain-specific design principles	7 experts different background team work		Observations, field notes	Interviews after course	Further guidelines, design patterns and tool support.
O.5	LL, sequence Hajanarvis [46]	N	EC, RS Logix 5000, RS View, market tool	Turn motors and valves Logic, sequence change including diagnosis and HMI LL (1), sequence (1) All approaches different sequence	Briefing sheets, questions, tool	45 skilled and 18 unskilled		Performance, Right first time rates, completion time, observation, sid effects of prior experience, Step logic and EC best results, EC superior for untrained Observations helped to identify problems the participants prevented from solving the task correctly	Self assed skill test	Observations revealed deficiencies in SFC programming interface
E1	UML 1.4 CD vs. ICL vs. PLC S7 UML 1.4 not restricted ICL truth table, SFC, idioms Friedrich *et al.* [69]-[71]	N	~ S7 IL, LL, FBD	p&p structure/behavior (hyb) DI 22, AI 1 DO 13, AO 3, WMC = 45 Modeling (1.5); PLC programming (1.5)	L (1.5), E (0.75) ppt-slides	BSc 5 (2 in each group) Students in Practice 3 Technicians 3	-	Behavior (19.31 steps from 32 realized $p < 0.01$) BSc 20.26, StiP 15.44, Techn. 12.78 ($p = 0.02$) Error rate: 43.84% ($p < 0.01$) BSc 53.36%, StiP 40.26% Techn. 13.64% ($p < 0.01$)	UML applicability structure 1.33, behavior 3.07 (1 not applicable-5 applicable) ICL (s 1.71; b 3.60) error: novices 48.7% experts 21.6% $p = 0.02$	Reduced number of UML diagrams, tool and modeling procedure needed
E2	UML PA (Instance structure D) vs. UML 2.0 (CD, Component D. Composite Struct. D, Deployment D) Focus: configuration, deployment Katzke *et al.* [72] [73]	N, M, P	Prot.	PE p&p structure (hyb, comm) ME: Continuous hydr. Press; switch between pressure and distance control, selection of pattern AI 60, AO 30 WMC = 3, UML (3.81); UML-PA (2.27)	PE L(3)/E(3) ME: L(3)/ E(-) 1 rep. online-help	BSc Mech. & CS 4th sem. 6	Q+ Restricted tool	PE: 20 min breaks, Too complex (difficult); ME: time to interpret UML 2.0 $\bar{t}_{20} = 22{:}52$, $SD = 3{:}14$ UML-PA $\bar{t}_{PA} = 13{:}40$, $SD = 4{:}42$ ($p \leq 0.05$) Self-assessment/steps UML 2.0 1.78, UML-PA 3.93 ($p \leq 0.10$) Errors [42]	~	• Tool flexibility needed • Change between diagram requires additional time • Communication and deployment support successful
E3	UML Embedded SC vs. SFC, focus: error handling and consistency check of sensor states Witsch *et al.* [75] [77]	N	Prot. CG	Cylinder (p&p) structure/behavior error handling: interception of time and erroneous sensor values; movement of 1 cylinder with two sensors DI/O 2(2)	L (1), E (0.75) per notation	BSc Mech. & CS 4th sem. 15	+ Small subtask	SC $\bar{x}$ 1.92 p/min ($\bar{x}$ 1.98 p/min with comp. states; $\bar{x}$ 1.81 without comp. states); SFC $\bar{x}$ 1.41 p/min	Subjects using comp. states higher prog. Experience: $r = 0.479$, $p = 0.033$	Error handling beneficial using UML SC with composite states Tool: side effects: demand for auto placing

E4a						BSc 2.Sem 41CD/21SC	-	no significant differences in terms of modeling order		
E4b	UML CD/SC vs. SC/CD order Vogel-Heuser, Seidel [78]	N	~	(p&p) structure/ behavior DI/O 13 (DI 9; DO 4), WMC = 11 small group (0.42)	L (3.2) E (0.42)	BSc 2.Sem 70CD/32SC	+	"behavior first" group: $\bar{x}$ 23.4 p (SD = 10.326; SE = 0.982), Structure first group $\bar{x}$ 18.4 p (SD = 8.220; SE = 1.825) out of 46 p max. s~: quality structural model bad: "structure first" $\bar{x}$ 12.26 p "behavior first" group $\bar{x}$ 11.14 p, structural quality of model is independent from modeling order	Post experiment interview and questionnaire to reveal challenges and reasons for mistakes ~	Forcing subjects to follow a specific modeling order is not helpful to improve structural models; subjects had problems to create suitable classes from similar objects
E5	UML Embedded CD/SC vs. IEC FBD Vogel-Heuser *et al.* [63]	N	ß CG	(p&p) *with reuse of buffer*, structure/ behavior DI/O 38 (DI 24; DO 14), WMC = 44 (UML) FBD 64 variables (2.2 + 0.92 quest.)	HLE and HB Fade out (Rep) (6.42)	18 Apprentices 1./2. year	++Strong correlation with abilities	Behavior (p = 0.22) and structure (p = 0.15) not significant; UML: thinking breaks Relatives to abilities (Table 1, 51): grade mathematics: (r = 0.234, p = 0.055); grade automation: (r = 0.327, p < 0.005) grade mechatronics: (r = 0.327, p < 0.005); cognitive demand: (r = −0.255, p < 0.05); previous knowledge: (r = 0.485, p < 0.001)	Frustration levels of UML group signif. higher than FBS group (p = 0.02), clearness of FBD signif. higher than of UML (p = 0.017) ease of use for behavior programming UML signif. lower than FBD (p = 0.012), Subjective quality estimation of UML group matches the factual quality signif. better than the FBD group (p = 0.025)	Difficulties with abstraction and building correct classes, e.g. cylinder, relationship CD and SC
E6	SysML-AT Parameter Diagram (PD) vs. CFC vs. IEC 61131-3 ST, Frank, Schutz/Obermeier, Schütz [25] [26]	N	~	Physical laws, structure, Maintenance task, analysis, understanding, interpret 4 - 5 sub-blocks, 7 - 8 variables, WMC = 5 - 6 (0.167 per notation + 0.083 quest.)	L (0.167) E (0.084) per notation	6 BSc. (mech. eng.) All notations different sequence	Q	Correct solutions PD 68.2% CFC 63.4% ST 64.3%	Answers correlated with objective results, learning effects across notations	Adequate method for requirement analysis and architecture design was missing
	SysML-AT vs. IEC 61131-3Notation + method Frank *et al.* [79] [80]	N, M			L (1.5) E (1.5)	15 BSc (mech.eng)	++	CFC $\bar{x}$ 93.5 (182) points NM $\bar{x}$ 123.1 (182) points (p < 0.001)		
E7	SysML-AT a) plus characteristics Frank *et al.* [81]	N, M, C	Prot.	NAS Coking plant structure, hyb, communication, RT: Material sorting, belt synchronization, level control; DO 3, AI 8; AO 4 WMC = 18, (a) 2,3; b) 2,6; c) 2,6); all mean times	L (2) E (2)	5	Q+	NM $\bar{x}$ 97.20 (182) NMC $\bar{x}$ 116.6 (182)	From NM to NMCP: mental demand and workload: ↓, fear of failure: ↑, fatalistic externality: ↑, NMC worst values: suitability for task and individualization (see Table 3)	Compromise between support and complexity. Difficulties with module abstraction
	SysML-AT b) plus pattern Eckert *et al.* [82]	N, M, C, P			L (2.5) E (2.5)	5	Q-	NMC $\bar{x}$ 116.6 (182) NMCP $\bar{x}$ 119.75 (182)		

Legend: N—notation; Prot—prototype; p&p—pick and place; L—Lecture; Q—only qualitative results; M—method; HB—handbook; (p&p)—part of p&p; E— exercise; er—error rate; ß—Version; A—analogue; I/O—input/output; Rep—repetition; CG—code-gen IEC 61131-3; P—pattern; D—digital; ~—only pencil & paper; C—characteristics; ME/PE—main/pre experiment; HLE—hybrid learning environment.

Table 2: Complexity values for heterogeneous modeling languages

Table Head	O	R	P	C(M)
UML Class Diagrams [44]	7	18	18	26.40
UML State Diagrams [44]	10	4	11	15.39
IEC 61131-3 FBD[1]	5	9	8	13.04
SysML-AT PAR[2]	8	2	9	12.21
CFC[3]	7	8	7	12.73
(SysML-AT a)[4] [45]	8	7	7	12.73

[1]The values listed in the table are gained by counting the following elements: O (function, function block, variable, (read, write), constant) = 5; R (direct connection 1 to 1, direct connection 1 to n, set, reset, rising edge, falling edge, negated connection, branch instrution, return) = 9; P (FB name, FB type, FC type, input name, output name, variable name, variable type, network

number) = 8; [2]O (Block, ConstraintBlock, attribute (In, Out, Local), PortIn, PortOut, Property) = 8; R (BindingConnection, FunctionalDependency; direct connection 1 to 1) = 2; P (BlockName, BlockEntityName, ConstraintName, AttributeName, AttributDataType, PortpertyName, PropertyDataType, PortName, PortDataType) = 9; [3]O (Function, Function Block, Input, Output, Comment, Composition, Selector) = 7; R (direct connection 1 to 1, direct connection 1 to n, Set, Reset, negated connection, branch instruction, connection marker based, return) = 8; P (FB name, FB type, FC type, input name, output name, jump mark name, execution sequence number) = 7;[4]O (Input Ports, Output Ports, Protocol Ports, Knots, Block, Functions, FBs, requirements) = 8; R (1 to 1 concrete, 1 to n concrete, n to 1 concrete, 1 to 1 discrete, 1 to n discrete, n to 1 discrete, enclosing) = 7; P (FB name, Function name, Requirement type, Requirement name, Knot name, Knot type, Blockname) = 7.

In prior experiments regarding process operators we tried to measure workload using a secondary task (e.g. communication and documentation) as, among others, proposed by Wickens [53] , but could not detect signifi- cant effects [54] .

Because for statistical significance a minimum of 15 subjects per different notation or approach, are necessary [55] , it is obviously impossible to conduct experiments with such a high number of experienced application en- gineers under same conditions.

In ergonomics it is usual to conduct usability experiments in engineering design with students of mechatronics or mechanical engineering because they are future application engineers. Maintenance tasks are performed in Germany mostly by skilled workers, therefore apprentices and technicians are appropriate subjects.

Training

Kim [56] suggests that the complexity of the design task in the training period and the test period should be in- creased stepwise.

As Kim and Lerch [56] , Ruocco [52] and others (i.e. [57] -[62]) point out, repetition is essential for learning object orientation. Ruocco decided for a stepped approach when teaching UML throughout a computer science program. He found that the application of UML during a database course and the incorporation of use case dia- grams, sequence diagrams and activity diagrams led to a richer and deeper exposure to UML.

In longitudinal studies, too, teaching beginners or freshmen in computer science or object orientation mostly goes along with repeated training [52] [57] -[62] . For training purposes, the pedagogic methods of repetition and fade-out, i.e. decreasing support by trainer from training step to training step, seem to be suitable (see for more details 63). In case of IEC 611131-3 or UML prior knowledge (see expertise in section subjects) acts as a dis- turbing factor if not equally distributed over subjects. Therefore training is necessary to adapt prior knowledge before conducting the experimental task. Time between training and experiment should be nearly the same for all subjects to avoid time depending differences in results.

Overview of Affected Variables (Usability Requirements)

In order to perform usability experiments, it is necessary to clarify how usability can be measured (affected va- riables) and which metrics can be used to make quantifiable statements about the advantageousness of the object of research (given by affecting variables).

The standard ISO 9241-11 [64] includes studies regarding use efficiency and the satisfaction of the users suggesting the measurement of product's usability in its context of use. According to this standard concerning usability requirements, the main affected variables, are (1) effectiveness, (2) efficiency and (3) user acceptance (cp. Figure 2).

Effectiveness, i.e. the quality of the result depends on the completeness and correctness of an engineered so- lution. Efficiency is the effectiveness in relation to the effort to engineer a solution. Both measures analyze the models developed by the subjects during the experiment compared with a so called master model [64] .

User satisfaction is the scale to which users are free of interference and their attitude to use a product [64] . Furthermore, the standard ISO 9241-110 contains dialogue principles for human-computer interaction as attri- butes to usability requirements. Those principles are suitability for task, for learning, for individualization, con- formity with user expectations, self-descriptiveness as well as controllability and error tolerance.

Effectiveness-Quality of the Resulting Model, Program

In (ISO 9241-11:1998) [64] it is proposed to determine the effectiveness by linking the grade of completeness with the grade of correctness.

(With N being the number of nodes, E the number of edges and R the number of errors with index task being the model of the subject and goal being the model of the experts taken as the correct solution in the master model).

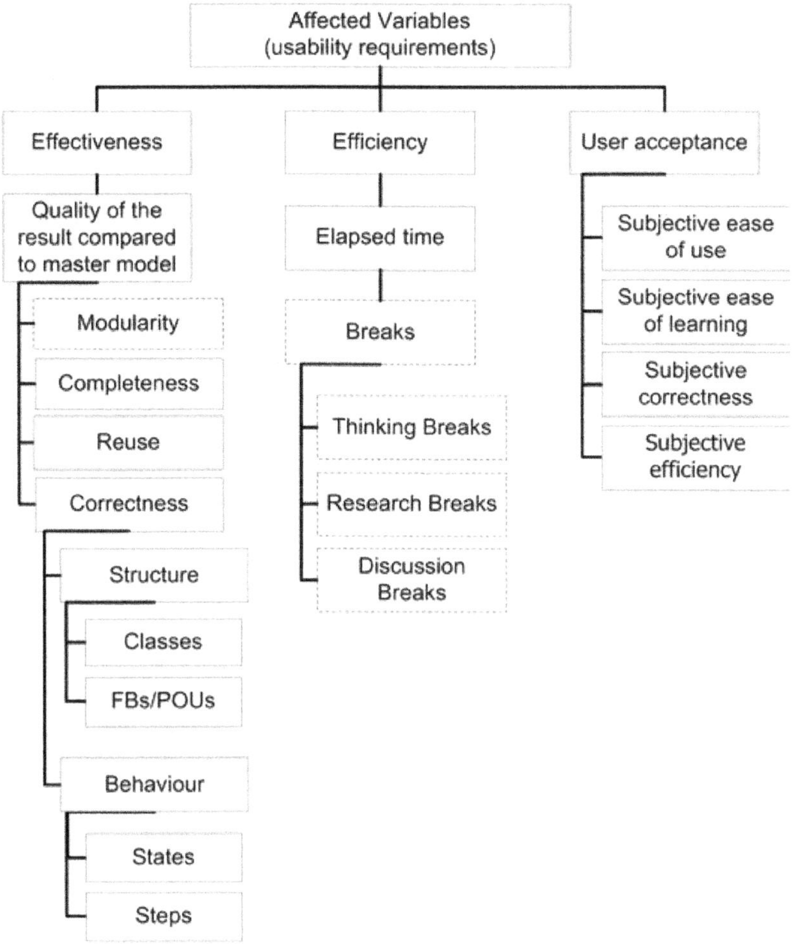

Figure 2: Usability requirements, i.e. variables in experiments.

Bevan (1995) [65] defined effectiveness in a different way as a product of quantity and quality:

$$\text{Grade of completeness} = \frac{\sum\left(N_{task} + E_{task}\right)}{\sum\left(N_{goals} + E_{goals}\right)},\ D = \{0;1\},\ W = [0,1] \tag{2}$$

$$\text{Grade of correctness} = \frac{\sum\left(N_{goals} + E_{goals} - R\right)}{\sum\left(N_{goals} + E_{goals}\right)},\ D = \{0;1\},\ W = [0,1] \tag{3}$$

Schalles compared only UML structure and behavior diagrams for business process modeling on an abstract level [44] . Applying his approach on MS would fall short. Using different notations modeled solutions are dif- ferent two regarding number of nodes and edges of the correct master model (see Appendix B for an example). As Strömmann [27] already realized often different correct solutions are created by different subjects, which should be evaluated equally good. Moreover, equality of nodes in a class diagram (abstract representation) and nodes and edges in an activity diagram (low level, object related) are rated equally by Schalles. In MS, the cor- rectness of structural model elements (e.g. classes or function blocks) should be measured differently than the correctness of low level behavioral model elements (e.g. correct steps or transitions from one state to another one [63]) due to the different degree of difficulty and ease of change in case of an error. Because IEC 61131-3 FBD is a language without nodes and edges Schalles approach is not feasible.

In accordance with Annett's proposal of an Hierarchical Task Analysis (HTA) [66] the proposed evaluation scheme counts referring to a top down approach all detailed elements modeled by the subject, allowing similar scores e.g. for a combination of class diagrams and created objects in comparison to structure elements in FBD.

In order to assess the effectiveness of notations the grade of task completion was used instead of measuring the grade of completeness and the grade of correctness separately before multiplying them. A task is completed, if its solution is logically and syntactically correct. As only correct task solutions, i.e. model elements are counted, it is not necessary to take additional errors into account. This results in the following term: (With N being the number of tasks).

$$\text{Effectiveness (F)} = \text{Grade of task completion} = \frac{\sum(T_{\text{solved}})}{\sum(T_{\text{goal}})},\ D = [0,1],\ W = [0,1] \tag{4}$$

This approach has three main advantages:

First, it can be applied equally for all kinds of notations as long as the given task can be completed with it and there is no restriction as in [65] , that only models with fewer nodes and edges than the master model can be evaluated.

Second, the task analysis can be used to select relevant tasks for evaluation and reduce the number of tasks to review, which then can be checked for logical and syntactical correctness, resulting in highly accurate data on task completion.

Third, no negative points for errors have to be used to calculate correctness, as this could manipulate results in an undesired way, e.g. errors lead to negative overall efficiency or errors in one part of the model nullify correct solution of others.

A fully automated analysis of the student's model compared to the master model is nearly impossible. The difficulty in rating the results of an experiment is comparable to a fair grading of exams by distributing points for correct solutions, but more sophisticated. As a consequence, the development of the correction guidelines for the manual evaluation is required. Points are given by two evaluators independently with a necessary interrater reliability of at least 65%.

Efficiency

Regarding industrial application the time needed to engineer an automation task correctly is one of the most important measures, defined as efficiency in usability evaluation. The efficiency of a notation can be calculated through a combination of effectiveness and time required for execution of a task (ISO 9241-11:1998).

Accoring to Schalles [44] , efficiency is defined as:

$$\text{Efficiency (G)} = \frac{F}{T} \quad (5)$$

(With effectiveness F and time T).

If time is a freely selectable variable, this calculation basically provides a good comparison between notations in terms of effect per time. For the experimental design time may be fixed and restricted to a calculated amount of time with GOMS [67] or pre-experiments similar to an exam or left open for subject's decision, delivering the modeling results when they feel ready. Fixed timing implies that the effectiveness measure already includes a statement on efficiency.

Nevertheless time should be recorded to allow analysis of modeling performance over time. Automatic sto- rage of the results in short time intervals allow, e.g. all 5 min or 7 min. the analysis of effectiveness per time in a more specific way.

User Acceptance-Subjective Aspects

Another possibly affected variable is the subjects' acceptance of the used notation (or tool) and the automation task when executing the task. Here, usability questionnaires based on the standard DIN EN ISO 9241 are best practice.

Moreover, aspects as the subjects' mental workload, control belief or motivation can be elicited (see Section 5, Section 4.7 for details).

For later analysis the affecting and affected variables and their relations will be evaluated to provide results for the comparison of the notation.

SELECTED USABILITY STUDIES

In the following Section usability studies (4.1) and usability experiments (4.2 - 4.8) with focus on different automation tasks are introduced and classified according to Section 3 (Table 1). The engineering task is classified as structural and/or behavior modeling task. The automation task's complexity is given by number and type of I/O, as well as weighted methods of class and number of variables in case of a classical PLC programming ap- proach using IEC 61131-3 FBD. The five related experiments are stronger related to case studies and to indus- trial application.

The seven experiments by the author's team (4.2 - 4.8) highlight different complex automation tasks and dif- ferent automation systems characteristics as given in Table 1.

Related Experiments

Experiment O.1 and O.2: Measuring Size and Complexity, Estimation of Development Time

Lucas and Tilbury and Lucas [67] [68] demonstrate how task analysis could be usefully applied for the prelimi- nary assessment of the effectiveness and perhaps even the efficiency of logic control design methodologies. Lu- cas [67] calculated the time to create a simple logic design program on the basis of low level user operations, e.g. keystrokes, mouse clicks and mental operations, for IEC 61131-3 Ladder Logic Diagrams (LL 405 min), Petri Nets (PN 1100 min) and modular Finite State machine logic (mFSM 1500 min) showing the significant differ- ence given by the notation itself. To derive the necessary steps and the used strategies, i.e. copy & paste, manual copy, they observed engineers during the design process and surveyed the time needed. Moreover, Lucas and Tilbury [33] [67] provide a way of comparing the complexity of control logic models respectively code of a simple lab scale MS created with the above mentioned notations plus SIPN by analyzing existing programs. They introduce quantitative measurements of complexity of a piece of code: size (i.e. number of operations and state variables), modularity (number of modules) and connectedness. Additionally, they introduce four typical scenarios for accessibility of data from a programmer's point of view, i.e. 1) single output debugging (specific questions regarding specific unexpected behavior in the machine), 2) system manipulation (how the user can manipulate the machine to achieve a desired state), 3) desired system behavior (desired behavior of the machine when examining only the schematics and the logic) and 4) unexpected system behavior (system's response to unexpected events). Because all these questions refer to already existing code they can be categorized to maintenance tasks. The four notations evaluated are compared regarding the four scenarios showing that Ladder Logic is still most appropriate for the first two but hard for Scenario 3 and 4, whereas Petri net, SIPN and mFSM are rated moderate or easy in Scenario 3, but minor in Scenario 1. LL is the small but very interconnected and mFSM the most modular, although largest program.

Experiment O.3 and O.4: Reusability Strategies

Strömman et al. [27] compared IEC 61499 with IEC 61131-3 in logic control design to foster reuse. Profession- als and researchers act as subjects programming a lifter application during a workshop. The resulting solutions differ totally showing different type of approaches, e.g. reuse of existing ST Code copied into an IEC 61499 frame, reuse of design patter, i.e. a state diagram, a mechatronic approach and classical IEC 61131 function block approach, concluding that guidelines to use IEC 61499 are required as well as an environment that fosters collaboration and exchange of information. The results were gained by model comparison and written feedback. Beforehand interviews were conducted to reveal the relevance of the study. Design approaches are context-de- pendent, i.e. the background of the designers, the existence of legacy software as well as business goals etc.

Based on this experience in experiment E0.4 Sierla et al. [28] organized one courses on IEC 61499 in 2005 to enable twenty practitioners and researchers to propose and negotiate about design alternatives in a team context with recorded interviews. In a second course in 2006 for professionals (3 subjects), researchers (3 subjects) and a standardization worker worked in a team representing the different social groups in a project evaluating the impact of team organization, knowledge integration, and software development method by an interview after the course. The benefit of a modular structure was realized as well as the risk of combining continuous control loops combined with sequential batch control logic. The necessity of shared guidelines, design patterns and tool sup- port was highlighted in more detail especially for batch control systems section.

Experiment O.5: Change of Sequence

Hajarnarvis et al. [46] compared 63 subjects applying for different methodologies changing the sequence of a given simple program, i.e. contact logic, step logic, SFC and EC. The participants had to change the sequence of a simple task with three motors and one valve. The authors identified different main problems, e.g. insufficient modifications for all but EC and incorrect algorithm for SFC and EC. The results are separated according to the participants' background, i.e. maintenance, planner, programmers and Rockwell personnel compared to the un- trained.

Experiment E1-Pure UML 1.4 and PLC Programming-Exploratory Study

The series of experiment E1 explored the influence of group work compared to individuals, the influence of prior experience in PLC programming and modeling, different qualified subjects, i.e. bachelor students of electrical and information engineering with students integrated into companies (StiP) and technicians modeling and programming a pick & place unit [69] -[71] (see Figure 3).

As affected variables the number of steps realized and their correctness was evaluated compared to a master model. The notations compared are UML, ICL and a control group only using S7 PLC programming languages IL, LL and FBD.

The results regarding quality of the model, i.e. error rates with 43.84% are disappointing. The high impact of qualification level on number of realized steps and errors is significant (see Table 1, E1 results). The influence of prior knowledge which is in this experiment only based on subjective rating in a questionnaire is evident, too. In this experiment prior knowledge leads to halve the errors. Subjects rate the applicability of both UML and ICL for modeling structural aspects as very poor and for behavior as fair. Comparing groups (2 subjects) with individuals, groups reach a higher number of modeled steps in (23.44 compared to individuals 15.04; $p = 0.01$), but unfortunately the error rate is not significantly reduced. The experimental results, e.g. the identified errors in the developed models (see 9, Figure B1) are used as input for the further development of UML for MS (E2, E3, E4, E5). The pure models and high error rate reveal an insufficient training and experimental design, but also the weakness of pure UML 1.4 as modeling notation. Subjects claimed a reduced number of diagrams with a clear procedure for UML modeling, a tool to support modeling with integrated code generation, because paper and pencil is not accepted.

Experiment E2-Deployment Using Pattern and UML-PA

Based on the results of E1 a domain specific language UML-PA was developed with a reduced number of dia- grams and domain specific stereotypes [72]. The research question was to prove the benefit of such a domain specific language under architectural aspects, i.e. regarding deployment of control loops and the related sensors and actuators connected via a field bus. The subjects should identify correct pattern and connect them to model the system from sensors to actuators including its deployment and communication relations. For this reason UML-PA provides ports to model communication interfaces in so called instance structure diagram.

The modeling approach using UML-PA and its instance structure diagram is compared with UML 2.0 dia- grams, i.e. class diagram, component diagram, composite structure diagram and deployment diagram. As auto- mation task a simplified real continuous hydraulic press was chosen with 30 control loops to be switched be- tween distance control and pressure control in case of overpressure. Each valve is equipped with a distance sen- sor to measure the valve opening and each control loop with a pressure transmitter. As additional input the press operator sets the set values of the pressure in the cylinder connected to the valve. The controllers output is the set value of the valve position and to the HMI the valve opening.

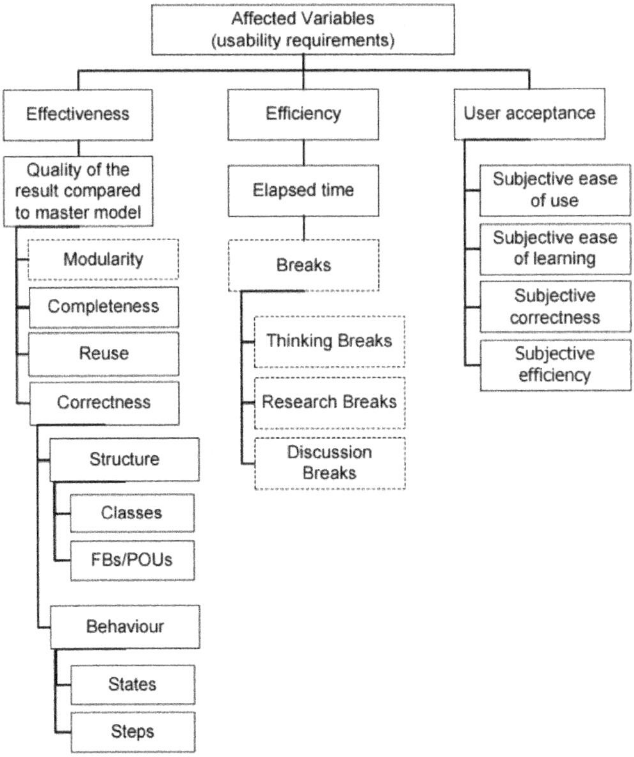

Figure 3: Pick & place unit (E1, E3, E4, E5).

UML participants checked their results after 1.78 changes and took the results as guidance to find an appropriate solution, UML-PA subjects checked their solution after 3.9 changes [72] (see also [73] for further information). The subjects properly analyzed the task and selected the given pattern establishing the required communication more efficient with UML-PA compared to UML 2.0 (see Table 1, E2), which is easy to understand due to the additional effort, i.e. diagram changes, needed in UML 2.0. This idea is included in the SysML-AT approach discussed in E7. The identified breaks and time needed to understand the relation between different diagrams needs to be optimized regarding improvement of MDE (see E5).

Subjects criticized the restricted tool. The restricted tool support encouraged students to follow a trial and error strategy which is unacceptable in a real industrial application.

Experiment E3-Error Handling Using plcUML SC vs. IEC 61131-3-SFC

Fulfilling the requirements from E1, a reduced number of diagrams with tool support and code generation, Witsch and Vogel-Heuser developed a prototypical plcUML editor implementing UML class diagram and state chart in a real IEC 61131-3 run time development with integrated code generation in CoDeSys 3.x [74] [75] (see also [1]). The plcUML diagrams are integrated similar to SFC as additional language transformed internally into a ST language derivate. Yang et al. [76] applied orthogonal regions in UML state charts to model primary sys- tem functions and corresponding traversal features and concurrent behavior. Witsch et al. [74] introduce compo- site states as groups of states allowing to model error behavior for those grouped states. Evaluation with experts showed the strength of the composite states for error handling as well as mode of operation, the focus of expe- riment E3.

The experiment validates that using state charts is more efficient than using classical SFC in IEC 61131 to proof cyclically sensor states regarding inconsistency as well as timing errors in a single moving cylinder, i.e. a cylinder component of the pick &place unit (cylinder in Figure 3). The mean steps programmed per minute us- ing state charts with composite states was 1.98 points/min compared to classical SFC in IEC 61131-3 with 1.41 points/min given the same points for both solutions to be reached [77] . The modeling speed of the SC group was significantly higher than the SFC group even if the SC subjects didn't use composite states.

The benefit of composite states is evident for error handling (see Figure 4, left), i.e. in SC the error handling for all states can be handled by one exception transition out of a composite state instead of multiple transitions, i.e. after each activity, error handling activities follow. If an error in the exception handling algorithm is identi- fied or an additional condition needs to be included modifications to the process can be covered in one path in SC, compared to multiple paths in SFC (cf. Figure 4, right).

Subjects using composites states estimate their programming experience higher than those who didn't use composites states. Many subjects criticized the absence of an automatic placement of elements in the tool, a site effect in the plcUML condition. In this experiment only exception handling was evaluated with a prototypical tool.

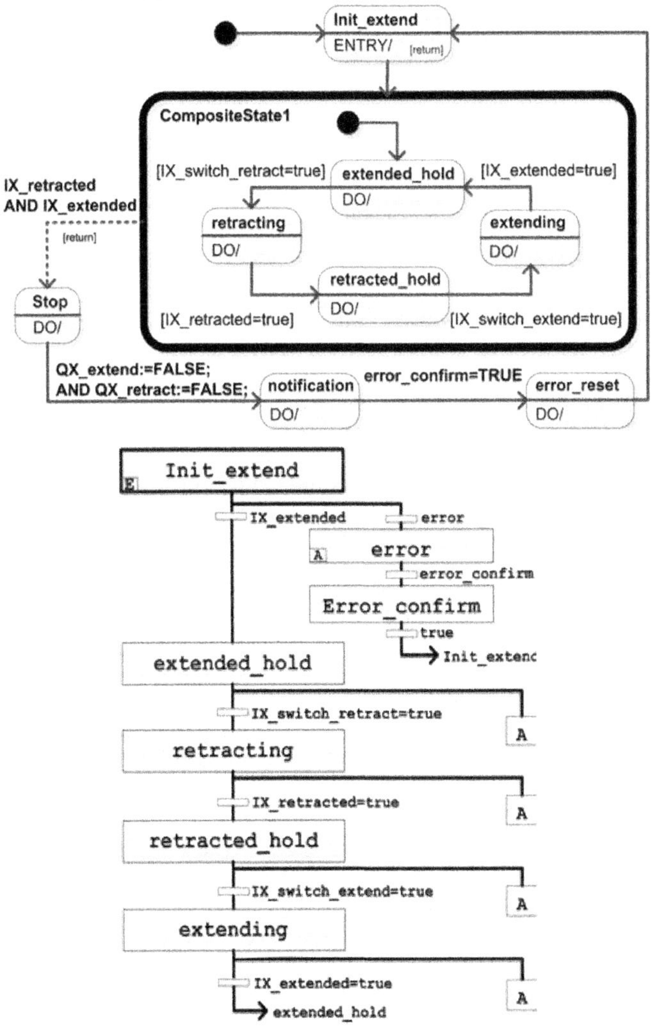

Figure 4: Comparison of subjects' best solution in SC group (left) and SFC group (right).

A more general design is discussed in Experiment E5 using plcUML with a more mature tool version.

Experiment E4-Sequence of Structure and Behavior Modeling in Workflow Using UML with Elaborated Training Concepts

The research question to be answered is, whether subjects can be successfully forced to model structure, when asking them to model structure before behavior or whether behavior first is a good strategy for engineers to achieve proper model quality. Therefore a training concept as well as a subsequent experiment has been devel- oped together with researchers from instruction theory [78] .

In a pre-experiment for E4 the main focus was to reveal whether the order of modeling is important for the quality of the model. The assumption was that students start with behavior modeling because it is easier for them, and then run short in time before finishing the structural model.

The pre-experiment was conducted without tool support only with paper and pencil after a training realized by a lecture and exercise in a very large classes (bachelor students 2nd semester mechanical engineering). The subjects were split in two groups: one group was told to start with structure modeling, the other with behavior mod- eling.

It showed that 35% of the subjects had problems to create suitable classes from similar objects of a plant in- cluding their attributes, and methods.

Examples of typical errors in the class diagram were (error rate in %):

· Objects were listed in addition to the classes, which inherit from the classes (23%).

· Classes were used in which objects of the class occur as attributes (7%).

· Single objects were modeled without classes (5%) [78] .

Overall, no significant differences concerning the modeling order could be found, but significant differences with respect to the trainer, as the two groups were trained by different teachers.

In order to eliminate that confounding effect in the main experiment one trainer trained for both groups. In that study, which has not yet been published, a larger sample (102 subjects) has been tested using the same pro- cedure and task as described for the pre-experiment above.

Here, the average participant reached 19.97 out of 46 points, i.e. lacked 26.03 points (SD = 9.1819). Regard- ing the performance measures, the "behavior first" group scored remarkably higher than the "structure first" group: While the participants in the "behavior first" group achieved 23.4 points on average (SD = 10.326; SE = 0.982), the mean value of the "structure first" group was only 18.4 out of 46 points (SD = 8.220; SE = 1.825).

In contrast, the structural modeling performance of the two groups was comparable (T = −0.972, df = 100, p = 0.33): participants in the "structure first" group achieved 12.26 points on average (SD = 5.029; SE = 0.601), while in the "behavior first" group they reached 11.14 points (SD = 6.067; SE = 1.072).

In the "structure first" group, the subjects reached only a mean of 6.14 out of 24 points in behavior modeling (SD = 5.083; SE = 0.607); in the "behavior first" group, however, the average behavior modeling performance was 12.25 points (SD = 6.112; SE = 1.080). This difference is highly significant (T = −5.278, df= 100, p = 0.00).

As a result for the next experiments we learned that the class room training was not suitable enough and that forcing students to follow a specific modeling order is not helpful to improve structural models.

Experiment E5-plcUMLvs IEC 61131-3 FBD with Apprentices Optimizing Training, Design and Analysis of Results-Exploratory Study

In this experiment the superiority of UML compared to FBD in a design task with a sophisticated training and with repetitive application of the notation, the ß-version of an UML tool (called plcUML), for a complex open loop control task and apprentices as subjects should be demonstrated. To allow further analysis between model- ing results and subjects' abilities and the development of an individual training fitting to individual abilities in a next step, selected abilities are collected as well as user acceptance.

As control task a sub-part of the pick & place unit with multiple reuse (only open loop control, weak real time requirements without communication requirements) should be modeled, i.e. three storage elements with one storage cylinder pushing the work pieces out of the storage and five different terminals with a terminal cylinder each, pushing the work pieces into the terminal. Because in industry very often skilled workers are conducting maintenance tasks and even easy design modifications 1st and 2nd year apprentices from a vocational school in Munich (89 subjects) act as subjects. Selected results of this experiment are reported already in [63] .

A hybrid learning environment (HLE), allowing to switching between computer-based and conventional in- structional designs] was developed and implemented. During training the groups repeatedly exercised program- ming and modeling tasks with increasing complexity (named fade out).

Several affecting variables related to abilities were obtained, i.e. grades in mathematics, German, automation, and mechatronics as well as cognitive capabilities, motivation levels, challenge, and workload (single instruments are described in [63]). As performance variable the programming/ modeling achievement was evaluated. To obtain this value, the developed models/programs were stored (every 5 min) and analyzed manually by two evaluators, who compared them to a master model. The subjects performance was measured as number of cor- rectly modeled or programmed elements and compared with respect to structure, e.g. classes or FBDs on the one hand and behavior, i.e. state charts and FBDs on the other (for details see Appendix A). Unfortunately, the re- sults were disappointing, because an overall significant benefit of plcUML compared to FBD could not be de- tected, but nevertheless interesting results could be found, e.g.

· OO modeling and FBD programming show different relations to variables like cognitive abilities, experience, workload, and knowledge the students' performance in the plcUML/CD + SC groups seems to be less re- lated to previous knowledge and cognitive abilities than students' performance in the 61131/FBD groups [63] .

· Subjects needed different times for structural modeling using UML/CD vs. FBD (see master modelFigure 5). Subjects needed in average 6.22 minutes more time for UML class creation (in comparison to the time needed to create the FB structure. This difference is slightly not significant (ANOVA, $F(1, 81) = 3.60$, $p = 0.06$), cf. [63] .

On the basis of the unexpected results further analysis of models, modeling process and the relation between model and results as well as subjective results have been conducted. Analyzing the main errors especially the errors in structural model, i.e. classes built:

- 42 subjects out of 44 built classes (including superfluous ones) as part of the structural model;
- 23 out of 42 used these classes in their behavioral model;
- 31 out of 42 modeled a second cylinder class separating storage cylinder and terminal cylinder, but 14 out of those 31 subjects built the second class identically besides the name, this indicates that they understood the class concept but use another type of abstraction, which is more related to the mechanical structure, i.e. a terminal and a storage cylinder are different instead of the software view in which both cylinders are identical.

Analyzing tool and training effects gathered from the subjective rating from questionnaire (Figure 6), 27 subjects of the plcUML grouped asked for additional training. From subjects' observation and analysis of the time needed, the authors expected the abstraction needed to build classes and the relationship between CD and SC to be the main challenge, because in the UML groups long thinking breaks occur before modeling classes. In the questionnaire only 5 subjects mentioned that development of classes is difficult, which is surprising regard- ing the above mentioned errors in building classes and the thinking breaks.

Regarding tool aspects (item 1.2 positive and item 2.5 negative in Figure 6) the plcUML tool seems to have some more problems.

Further subjective results gained from the questionnaires were: 1) Frustration levels were significantly higher in the UML group compared to the FBD group ($p = 0.02$); 2) The clearness of FBD was rated significantly higher than of UML ($p = 0.017$); 3) Behavior programming was rated significantly easier with FBD than with

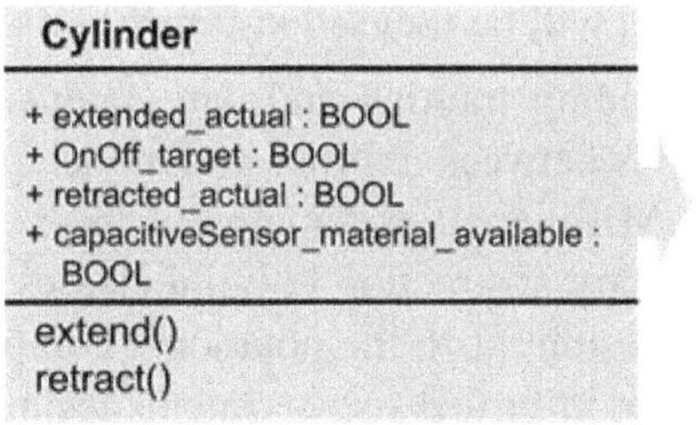


```
PROGRAM Mainprogram
VAR_INPUT
    Cylinder1:Cylinder;
    Cylinder2:Cylinder;
    Cylinder3:Cylinder;
    Cylinder4:Cylinder;
    Cylinder5:Cylinder;
StorageCylinder1:Cylinder;
StorageCylinder2:Cylinder;
StorageCylinder3:Cylinder;
END_VAR
VAR_OUTPUT

END_VAR
```

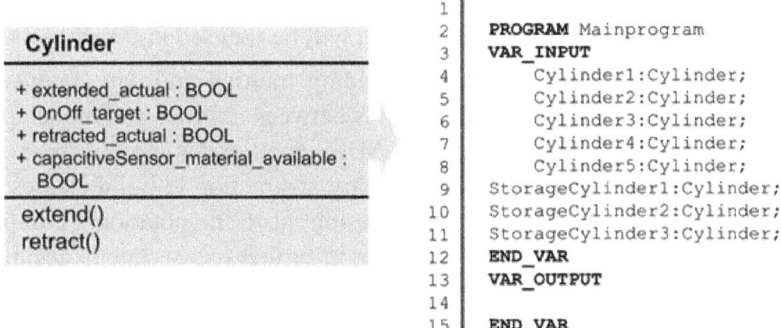

Figure 5: plcUML class diagram master model integrating storage and terminal cylinder and its IEC representation.

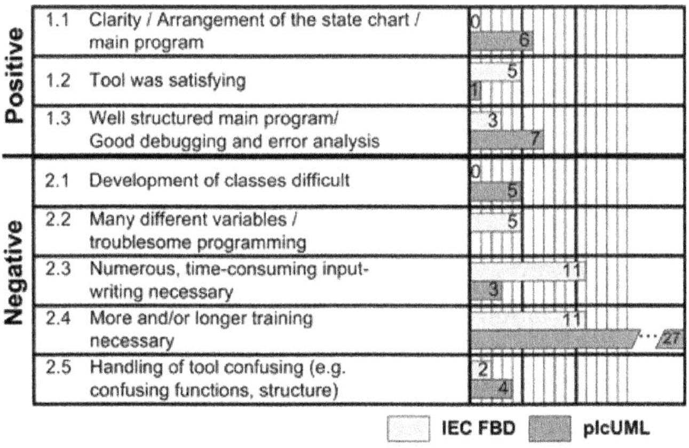

Figure 6: Subjective statements after the experiment.

UML (p = 0.012). And 4) subjective quality estimation and factual quality match far better with UML than with FBD (p = 0.025).

Because of the observed thinking breaks, we analyzed the modeling progress over time (points over time) in a random sample of only three subjects (with similar quality of model) we found differences in plcUML and IEC group, in the plcUML group there is a longer period of time until points referring to the master model are ga- thered and there is a clear ramp in points compared to the more steady increase of points in FBS group (Figure 7).

For further experiments detailed analysis of modeling progress is needed and, therefore, the cycle time of storing data needs to be reduced and a more efficient approach of analyzing subjects' modeling process over time needs to be developed. Details for the analysis and rating of subjects' models compared to the master mod- el are given in Appendix A.

Subjects debugged at different times, some at the beginning and others at the end of the experiments with a nearly complete model. The analysis of debugging is necessary to find errors and will be focused in future work.

The design of the experiment including training and data analysis was appropriate delivering detailed rela- tions between abilities and model quality, but revealing still shortcomings of plcUML as notation for apprentices in design tasks. Our assumption is that the necessary abstraction to build classes is too high for this group of subjects. These results fit to the notational complexity of class diagrams of Schalles. Therefore, in further expe- riments technicians and engineers will be included as subjects with a higher level of knowledge and experience in PLC programming. Additionally, different levels of task complexity will be tested.

Experiment E6-Maintenance Task in Early Phases of Notation Development with SysML-AT vs. Continuous Function Chart (CFC)

The research question of experiment E6 is how to evaluate three notations in a qualitative way in a very short period of time for training and experiment in an early phase of the development of a notation. E6 evaluates dif- ferent modeling and programming notations (see also [16]), i.e. Parametric diagram (PD) of SysML-AT [26] vs. Continuous Function Chart (CFC) and IEC 61131-3 Structured Text (ST), regarding a maintenance task, i.e. understandability (analysis and interpretation according to [31]) of model contents in a qualitative way. The ex- periment was based on three different simple models of physical laws (about 4 - 5 sub-blocks and 7 - 8 va- riables), with each model described in every considered notation. Because the evaluation should take place in an early design phase and the time needed should be very short, tool support is not applicable. Bachelor students of mechanical engineering worked without a tool after a very short training, passing all three different notations and all three models. The sequence of the notations was permuted for each subject (see Figure 8) to eliminate learning effects.

As a software maintenance scenario, the subjects had to correctly interpret the models' contents, consisting of components (sub-blocks and variables) and data flows to answer questions regarding the model contents cor- rectly. The mean of correctly answered questions was highest for the PD (68.25%) with a positive offset of 3.97% to ST (64.28%) and 4.76% to CFC (63.49%) (see Table 1, E6). The experiment shows, that even a short training with a short time for experiment and a small number of subjects delivers qualitative results. In accordance with the results, in questionnaires that tested the subjective cognitive demand, the subjects rated the PD as the most understandable notation. Furthermore, all of the subjects answered, that they experienced a learning effect re- gardless of the different notations they used.

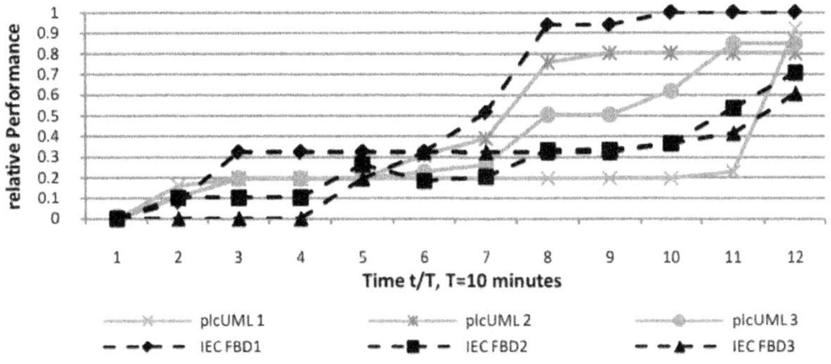

Figure 7: Modeling progress over time for 3 subjects of each group

Subject 1	Subject 2	Subject 3	Subject 4	Subject 5	Subject 6
Previous Knowledge Tests & Trainings					
Model A CFC	Model A CFC	Model A PD	Model A PD	Model A ST	Model A ST
Model B ST	Model B PD	Model B ST	Model B CFC	Model B PD	Model B CFC
Model C PD	Model C ST	Model C CFC	Model C ST	Model C CFC	Model C PD
Questionnaires for subjective cognitive demand & discussion					

Figure 8: Experimental design of E6-maintenance task.

The results of a focus group that was conducted for additionally evaluating the SysML-AT [25] also indicated that the developed modeling approach is well suited for automation software modeling.

Experiment E7-Conceptual Engineering of Structural aspects of Distributed Networked Automation Systems (NAS)

The research question in this experiment is whether additional support in structural modeling of NAS realized with characteristics and pattern is beneficial in the conceptual design or whether the resulting complexity hinders the benefit. Besides the instruments regarding user acceptance are more elaborated and should give answers in more detail in relation to quality of models.

As the results of E5 and E6 show, plcUML and SysML-AT have positive influence on the programming of a PLC. Following Sierla [28] and the difficulties identified for engineering of distributed systems E7 evaluates a SysML-AT based notation and workflow vs. CFC for a high-level design of NAS in MS (see also [16] [45]). The evaluated approach focuses on the overall design of NAS integrating notation SysML-AT being the successor of plcUML. The SysML-AT based concept contains a workflow procedure referring to a life cycle model (following requirements E1) including communication and real time requirements for a hybrid control task (experiment E7 a). Additionally, characteristics (E7 b) and characteristics plus patterns in (E7 c) are compared to the pure notation and workflow procedure. Conditions b and c are only qualitative measures, because of the small group size.

The approach covers the modeling of automation hardware and software as well as of functional and non- functional requirements. From described requirements the functions that need to be implemented can be derived and captured within the same model. Hardware elements like sensors, actuators and nodes and their interfaces and properties are considered within the modeling approach as well. This enables the integration and linking of hardware and software models [79] . The notation is based on the SysML Block and Requirements diagram using ports to represent software and hardware interfaces (Refer also to [80] . Duration of experiment was not re- stricted, but taken as measure (mean given in Table 1, E7).

Characteristics as well as pattern supported subjects in solving the task, i.e. design of the automation concept of a coking plant including belt synchronization without implementation. Main task of the experiment was to conceptually design a closed loop for speed synchronization which included three belts. This comprised all ne- cessary functions, interfaces and relations to the sensors and actuators. The internal behavior and control algo- rithms were not required, i.e. the structural part of the model needs to be designed. Characteristics detail re- quirements as well as the later design solution including element relations. During the design the comparison of requirement characteristics and solution characteristics help to decide if the solution fits the requirements (cp. Figure 9).

Additionally patterns, divided into functional and deployment patterns, help to find a solution. Functional patterns include proposals and support the engineer in the development of the functional model. Deployment patterns indicate distribution alternatives of functions and support the engineer in the development of the deployment model [82] .

The models subjects stored after finishing the given task were analyzed compared to a master model. The re- sults show major difference between the subjects' solutions and master model, i.e. experts' best practice regarding module structure. Similar to Sierla [28] different possible solutions were detected, i.e. most subjects chose a functional oriented modeling approach, instead of a mechatronic approach taking modularity, reuse and architectural aspects of NAS into account. The experiment intended that students follow a mechatronic approach, therefore the master model was built realizing the mechatronic paradigm. In further experiments either the me- chatronic approach needs to be integrated into the training or subjects' mental models need to be collected before- hand.

Nevertheless, the experiments show a significant benefit of SysML-AT compared to CFC (see Table 1, E7 a to c column results). Regarding notation with life cycle model, i.e. NM subjects gained significant better models compared to CFC (123.1 mean compared to 182 max. points, the best subject gaining 144 points, see appendix C). Using characteristics additionally (NMC), subjects improved their models again. But for those experiments only qualitative results are available due to the limited number of five subjects. Regarding user acceptance measures subjects stated less mental demand using pattern (see Table 3), i.e. NMCP has lowest mental demand with 10.75 of max 20 points. (the higher the more mental demand). The motivational factor "fear of failure" was most pronounced in the group with patterns and characteristics (NMCP). Furthermore, this group showed high external control beliefs meaning that subjects strongly related the outcome of the results to external circumstances and high fatalistic externality meaning that success is assessed as depending on fate, fortune, and chance and, however, subjects perceived low mental demand during task performance. In addition, according to usability aspects, suitability for task was best evaluated for group with characteristics (NMC). Suitability for individualization of patterns was rated significant lower than both other conditions.

Based on UML-PA and E2 as well as the experiences gained and rules derived for E5 (including task development, training and tool development) and experimental design in general (see 6) the experimental design of E7 was developed appropriately evaluating also the derived rules (see 6). E7 evaluated the benefit of NM for NAS and hybrid control with real time and communication requirements. Even relations to abilities realized in E5 could be further developed with a more advanced questionnaire. Results reveal more relations to human factors, e.g. mental workload, and usability measures. For further engineering support the challenge is to find a compromise between supports by characteristics and pattern and the approaches' complexity.

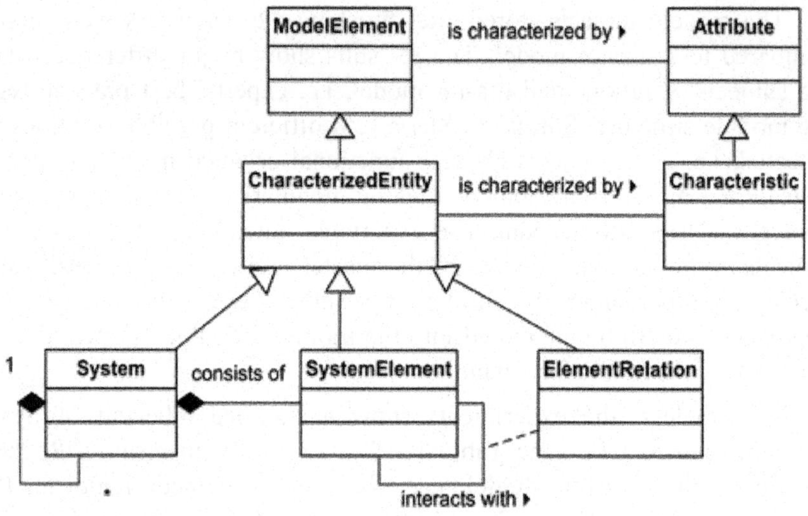

Figure 9: Characteristics meta-model [81] .

Table 3: Results of human factors and usability measurement in E7

Variable	Mean			
	NM	NMC	NMCP	p
Motivation				
Fear of failure	14.26	15.00	23.67	$p < 0.05$
Control beliefs				
Fatalistic externality	45.95	55.20	60.75	$p < 0.05$
Mental workload				
Mental demand	14.28	14.80	10.75	$p < 0.05$
Attributes for usability requirements				
Suitability for task	27.35	30.00	20.75	$p < 0.055$
Suitability for individualization	26.59	26.00	13.75	$p < 0.001$

Summary of the Experiments

All experiments focus on the design phase besides E6, a centralized single PLC as control hardware besides E7 and students as subjects besides E1 and E5.

- E1 was the first experiment exploring the method of usability evaluation in logic design engineering with a single closed loop controller and compared pure UML 1.4 and PLC in a first attempt without the support of an engineering tool and with a large unstructured task.
- E2 focused on a hybrid automation task including communication with the focus to support deployment by simple UML-PA pattern compared to classical UML 2.0 with restricted tool support and a narrow engineering task.
- E3 focuses on error handling comparing plcUML State Chart to Sequential Function Chart (SFC) in IEC us- ing a very simple automation sub-task and a short classical training.
- E4 focuses on SC vs. IEC 61131-3-SFC sequence of structure and behavior modeling in workflow using UML 2.0 with a didactically more elaborated but classically conducted training concepts in smaller sub groups with the goal to increase the quality of the structure model.
- E5 is similar to E1 also an exploratory experiment further developing the method of usability engineering experiments using a real software engineering tool with embedded UML the so called plcUML compared to IEC 61131-3 FBD with apprentices optimizing repetitive training, and exercise, an elaborate training envi- ronment and smaller automation task with reusable sub-process, including also human factors and prior knowledge.
- E6 focuses on a maintenance task in the early phases of notation development with SysML-AT vs. Continuous Function Chart (CFC) to show the benefit of easy and quick sub-experiments in the development pro- cess of the notation.
- E7 Conceptual Engineering of Structural aspects of distributed networked automation systems (NAS) in- cluding a procedure for life cycle support and characteristics for pattern selection and reuse with a detailed analysis of user acceptance including motivation.

In every single description of an experiment the research questions as well as the most important aspects of the experimental design, results and lessons learned regarding usability aspects are discussed as well as results for further development of MDE, i.e. notation, procedure and tool. Most experiments are based on prior experi- ments and notational development resulting from a prior experiment is tested in one of the following experi- ments.

SELECTED RESULTS FOR FUTURE USABILITY EXPERI- MENTS

The following section summarizes the best practice rules gathered to the best of our knowledge. At first the criteria for the selection and configuration of the affecting variables (see Figure 1) are discussed, e.g. the task, the training and selecting a group of subjects. Afterwards the criteria for selecting the affected variables are dis- cussed (see Figure 2).

Configuration of Affected Variables

Task Development

As affecting variable (see Figure 1), the type of the engineering task (maintenance or design) and automation task complexity and characteristics are key issues in relation to the complexity of the new notation or approach to be evaluated and the time available for training and the experiment itself.

1) Automation Task

To classify or rank the automation task complexity compared to other experiments and to estimate the time needed for training as well as the task itself in the experiment, the authors introduced some measures, i.e. num- ber of I/O, number and type of control loops and depending on the used notation the WMC and number of states for OO design and the number of FBDs and variables for classical PLC programming using IEC 61131-3. Be- sides the task characteristics, i.e. real time, communication requirements and the tasks type as well as the inclu- sion of exception handling (E3) and mode of operation are relevant, too. In the above introduced experiments the WMC reaches from 3 in E2 a strongly restricted experiment using pattern to 43 in E5 and 45 in E1 in a more industrial related scenario.

It is obvious that a complete engineering task consists of a lot of decision points with different ways to a cor- rect solution. These variation possibilities need to be covered by an evaluation scheme.

2) Engineering Task

Starting with HTA or GOMS, the required steps to fulfill the task are found. The quality of the HTA depends on the skills and experience (also industrial) of the experts conducting the HTA. Interviews with industrial experts are helpful to find appropriate subtasks as well as typical module libraries available to be provided in the experimental setup.

Modeling mostly consists of structural and behavioral aspects. In most of the experiments described above, structure and behavior were an issue (Table 1, column engineering task). All experiments besides E6 dealt with design and model creation (E2 model configuration). E6 highlights maintenance tasks and showed that tool support may be neglected for easy tasks as well as training may be very short compared to design tasks.

For both, modeling and training, the designer has to decide whether to provide a life cycle model or even a method and a tool. For more complex engineering tasks a tool is a prerequisite to gain subjects acceptance (not reached in E1, E4 and E6) and motivation. On the other hand, a prototypical tool (E3) leads to results that may be induced by the tool and not by the notation to be evaluated. Sophisticated tools need additional time for training. The prototype plcUML or SysML-AT, therefore, needs to be carefully tested by novices and persons belonging to the qualification group of future users before conducting the experiment, to ensure an effective de- tection of as many defects of the tool as possible prior to the experiment. Since otherwise frustration will rise and may act as disturbing factor in the experiment (E7 c).

Development of Training

As discussed above, an appropriate training is a prerequisite for meaningful results, but hard to achieve (E5 not E7 c)) in the first experiments. A hybrid learning environment is advantageous to reduce disturbances by indi- vidual trainers as in the pre experiment of E4. Furthermore, process simulation offers high benefits as to testing and debugging the software. For more complex notations and procedures, e.g. OO and UML, repetitive training with fade out is beneficial (E5). A training period of 1.5 days for OO with apprentices as subjects and 0.5 days for E7 a) with students as subjects was appropriate. With a very simple task or strictly focused hypothesis and a restrictive tool significantly shorter duration can be reached (E2 and E6).

Selection of Subjects

Besides E1, we decided for individual subjects to allow the identification of reasons and dependencies to indi- vidual abilities. This excludes to examine benefits of group work as found in Sierla [28] . In the field of MS en- gineering students are a typical group of subjects for design tasks as well as technicians and apprentices for maintenance tasks and simple design modifications at customer site. The necessary numbers of subjects per cell to gain quantitative results is minimum 15. Different skills and abilities, e.g. mathematics are often related to results and act as disturbance factors.

Pre-tests are recommended to adjust distribution of subjects to groups regarding expertise and abilities. Dif- ferent tests are available (E5) or adaptable (e.g. on general intelligence [83] or on previous knowledge). Missing or insufficient motivation may also be a disturbing factor as realized in experiment E7c. Also, mental workload, i.e. the cognitive demand perceived during modeling tasks is a critical factor for the probability of errors and, therefore, should be at an intermediate level (E5 and E7). When analyzing specific aspects of a notation in more detail after the main experiment, group sizes from 6 to 8 are regularly implemented to get qualitative results. In E6 the sequence of the notations was permuted for each subject to eliminate learning effects instead of using one notation for one group, which in case of E6 would have multiplied the necessary number of subjects by three.

Measuring Affected Variables/Usability Requirements

To analyze the gained result and to evaluate it, master models are recommended, developed by the designer of the experiment together with other experts.

Data Collection-Organizational and Technical Challenges

For the data analysis observation and recording of subjects' results are most important. The easiest way to observe subjects is to take a video, but the manual analysis of the video is time consuming. In engineering tasks using an engineering tool, the most often implemented strategy is to store the model cyclically with a selected time (all 2 or 5 minutes E5 and E7 5 min) or if a new input is typed in the model (E2). The cyclical storing strategy has the disadvantage of losing information in between storing intervals similar to the sampling of an analogue value. Storing the model with every subject's input has the disadvantage of large amounts of data, which need to be analyzed later. The strategy may not be integrated in real tools as necessary if using a ß-version of an industrial tool (in E5). The challenge is to implement storing strategies in the prototype or to get access to a market leading tool in case it should be used for evaluation. The CoDeSys implementation was easy to realize for the authors' team due to the gathered developer's knowledge of the plcUML-Plugin. Additionally to model analysis, human observers are advantageous especially in case of pre-experiments and to include additional information gathered by observation. Unfortunately, this is expensive, because the observers need to be trained; the observation needs to be documented in a standardized form and approximately 1 observer is required for 2 - 4 subjects. In E5 long periods of thinking breaks in the OO groups before building classes were found and included in further analysis. The analysis of results gained, i.e. model consolidation over time seems to be useful, but is depending on the availability of data and ease of analysis.

In psychology thinking aloud is an often implemented method, which is often not accepted and applicable by engineering students (E1). Another issue is to gain information why subjects make mistakes or chose a specific solution. To a certain degree this information may be gained by individual interviews or online questionnaires directly after the experiment (E4). In E4 subjects were asked to analyze their solutions compared with the master solution and give reasons, e.g. lack of time, translation problems, distraction etc., for their mistakes. The method is promising, but hard to realize with large groups of subjects because of possible interviewer effects with regard to the questions asked.

Effectiveness

Usability evaluation concerning affected variables, i.e. effectiveness, efficiency and user acceptance was realized with different methods. To assess effectiveness, completeness and correctness are measured by counting the numbers of correct steps compared to the master model, e.g. in the behavior model, e.g. a state chart the number of steps, in the structure model in FBD the number of variables, the number of classes and objects in a class diagram (for evaluation scheme E5 see Appendix A).

The difficulty in rating the results of an experiment is comparable to grading exams by distributing points for correct solutions, but more sophisticated. Points are given by two evaluators independently with a necessary interpreter reliability of at least 65% (E5, E7 see Appendix).

Efficiency

To evaluate efficiency time stamps need to be included in the stored data and analyzed or as mentioned in 1) the cyclically stored data are taken to analyze efficiency over time. In most experiments efficiency is effectiveness in the given period of time subjects got for the experiment. In most experiments time was limited due to organi- zational reasons, besides E7 where time was taken as a variable: When subjects felt ready they submitted their solution and the time needed was stored.

User Acceptance

For evaluation of user acceptance in all of the above described experiments questionnaires based inter alia on the EN ISO 9241 and on recognized tests as

RSME [83] and NASA-TLX [84] were implemented and further de- veloped from one to the next experiment to analyze subjective values regarding modeling as such, the notation evaluated and/or the tool used, e.g. E1 and E5. Furthermore, extended evaluation of attributes for usability re- quirements examined by EN ISO 9241-110 questionnaire (E7) was additionally used to collect users' assess- ment of applicability of patterns and characteristics. Results revealed suitability for task and for individualiza- tion as appropriate indicators of difference. Questionnaires regarding the notation and tool may also reveal weak- nesses of training and notation (E5 class concept).

SELECTED RESULTS FOR THE DEVELOPMENT OF FU- TURE NOTATIONS FOR MODEL BASED SOFTWARE ENGINEERING

In MS, hybrid control tasks, real time and communication requirements of different complexity need to be engi- neered during design and maintained during operation covering structure and behavior in MS models.

From the results of E1, we realized that pure UML 1.4 with its five diagrams used in E1 is confusing and not appropriate especially for structure models. Additionally embedded tool support in PLC development environ- ments and a procedure is requested by subjects. Forcing students to follow a specific modeling order, e.g. beha- vior or structure first (E4) is not helpful to improve structural models. The introduction of plcUML embedding class diagrams and state charts into an IEC 61131-3 tool enlarged with composite states for error handling showed benefit, but tool aspects as placement were criticized (E3). In E5 a more general, but simple logic design task with reuse revealed weaknesses of plcUML in design tasks for apprentices using a ß-Version of the tool. The challenge for apprentices was the necessary abstraction when building classes. Weaknesses in training and tool were criticized (Figure 6). The tool has been further developed and integrated in CoDeSys by industry in June 2013 now used in different industrial companies and research. Experiments focusing on maintenance tasks, evaluated in E6 with students of mechanical engineering, indicated that the SysML-AT PD has advantages compared to CFC and ST (qualitatively).

All these evaluations concentrated on the automation software of one centralized PLC. Regarding deployment and NAS two experiments were conducted, i.e. E2 and E7 including communication and real time requirements. In E2 a domain specific UML the UML-PA with reduced number of diagrams was beneficial in deployment of software to hardware devices like PLCs, using patterns with a very simple conceptual control task. The restricted tool was criticized, but the reduced number of diagrams was advantageous compared to UML 2.0. plcUML, consists of the Class Diagram for modeling software structure as well as the Activity Diagram and State Chart for modeling discrete software behavior using Activity Diagrams in the early phases of the software lifecycle for specification issues and the State Chart for detailed modeling of behaviour. Further developments of plcUML, namely the SysML-AT added the SysML Parametric Diagram for modeling constraints as mathematical equations to describe physical laws to the diagrams of plcUML. Although advantages of both notations were noticed, the results from E6 (focus group) indicate that a MDE approach for MS has to consider and support requirements analysis and architectural design and a supporting method. Especially for NAS the architectural design is even more important. Such a method was developed and positively evaluated in experiment E7 to be most ap- propriate for all typical requirements of automation in MS. Recent works currently develop an approach that contains the developed methodology for NAS and requirements modeling followed by software modeling and generation based on the plcUML and SysML-AT.

CONCLUSION AND OUTLOOK

MDE approaches should increase efficiency and quality in design and maintenance of software engineering for MS. The article showed results of usability experiments using pure UML, domain specific UML versions, i.e. UML-PA and UML E as well as domain specific SysML-AT for mainteance purposes and NAS. Summarizing the most important technical issues pure UML 1.4 or 2.0 is not appropriate, but plcUML with reduced number of diagramms and a supporting modeling process integrated in an IEC 61131-environment to support roundtrip engineering. For error handling plcUML SC with commmposite states are beneficial compared to IEC 61131-3 FBD. Structural modeling using pure or even plcUML is still a challenge for many subjects as well as the creation of classes in the sense of abstraction used in computer science. Abstraction in automation and mechatronics is different to computer science, i.e. more related to physics, also in distributed systems application. Complexity of notation (class diagramm and E7) relates to difficulties in applying the notation in an experiment with time restrictions (2 days). For NAS the applicability of notation was positively and quantitaive evaluated and for characteristics and pattern further experiments and longer training time is needed. Ongoing research is looking at a detailed analysis of humans' mistakes trying to find reasons by interviewing subjects after the experiment.

Regarding real industrial software engineering tasks in MS all these experiments lack of experienced subjects, i.e. application engineers and the start-up phase with debugging. Real applications and some applications engineers are included in [85] . The classical debugging phase to find faults is not explicitly analyzed up to now even if Myers [86] provides an interesting approach to classify runtime faults and the underlying software errors. Debugging in E5 was limited to simulation and restricted due to given time. At the moment we implement inte- views after another experiment focusing on reuse of modules with apprentices to analyse faults categorized to Myers' classification.

Regarding usability aspects the presented experiments proofed the relevant affecting and affected variables (Figure 1 and Figure 2) to be taken into account when designing the experiment.

To increase efficiency and quality of software in the development process of an industrial company in ma- chine and plant manufacturing model based approaches using notations as UML and SysML are applicable and could be proven as partially quantitive beneficial. The prerequiste for a real benefit is the availability of an inte- grated tool support in the IEC 61131-3 especially for maintenance reasons to guarantee consistency of model and implemented code. Nevertheless, it is will not be easy to introduce and implement MDE using UML and SysML in an industrial company. Training and rules for application are necessary as well as a workflow to inte- grate existing legacy software developed in years. To integrate legacy software the existing software needs to be analyzed at first and modularity concepts need to be developed as a prerequisite for MDE. Variablilty analysis from software engineering should be implemented to maintain and evolve models and code synchronously.

Further research is also needed regarding the integration of more advanced controllers into the usability eval- uation, e.g. modeled in Matlab/Simulink.

ACKNOWLEDGEMENTS

The author gratefully acknowledges the support of the German Research Foundation (DFG) for the projects DisPA (Vo 937/2-1), KREAagentuse (VO 937/8-1) and FAVA (VO 937/13-1) and the support and fruitful dis- cussions with Christoph Legat, Daniel Schütz, Kerstin Duschl, and Martin Obermeier.

REFERENCES

1. Basile, F., Chiaccio, P. and Gerbasio, D. (2012) On the Implementation of Industrial Automation Systems Based on PLC. IEEE Transactions on Automation Science and Engineering, 10, 990-1003. http://dx.doi.org/10.1109/TASE.2012.2226578

2. International Electrotechnical Commission (2013) IEC International Standard IEC 61131-3: Programmable Logic Controllers, Part 3: Programming Languages. IEC, Geneva.
3. Thramboulidis, K. (2010) The 3+1 SysML View-Model in Model Integrated Mechatronics. Journal of Software Engineering & Applications, 3, 109-118.http://dx.doi.org/10.4236/jsea.2010.32014
4. Rzevski, G. (2003) On Conceptual Design of Intelligent Mechatronic Systems. Mechatronics, 13, 1029-1044. http://dx.doi.org/10.1016/S0957-4158(03)00041-2
5. Zhabelova, G. and Vyatkin, V. (2012) Multiagent Smart Grid Automation Architecture Based on IEC 61850/61499 Intelligent Logical Nodes. IEEE Transactions on Industrial Electronics, 59, 2351-2362. http://dx.doi.org/10.1109/TIE.2011.2167891
6. Sauter, T. and Lobashov, M. (2011) End-to-End Communication Architecture for Smart Grids. IEEE Transactions on Industrial Electronics, 58, 1218-1228.http://dx.doi.org/10.1109/TIE.2010.2070771
7. Vyatkin, V. (2013) Software Engineering in Industrial Automation: State-of-the-Art Review. IEEE Transactions on Industrial Informatics, 9, 1234-1249.http://dx.doi.org/10.1109/TII.2013.2258165
8. Yang, C. and Vyatkin, V. (2012) Transformation of Simulink Models to IEC 61499 Function Blocks for Verification of Distributed Control Systems. Control Engineering Practice, 20, 1259-1269. http://dx.doi.org/10.1016/j.conengprac.2012.06.008
9. Dubinin, V., Vyatkin, V. and Pfeiffer, T. (2005) Engineering of Validatable Automation Systems Based on an Extension of UML Combined with Function Blocks of IEC 61499. IEEE International Conference on Robotics and Automation (ICRA), Barcelona, 18-22 April 2005, 3996-4001.
10. Secchi, C., Bonfé, M. and Fantuzzi, C. (2007) On the Use of UML for Modeling Mechatronic Systems. IEEE Transactions on Automation Science and Engineering, 4, 105-113. http://dx.doi.org/10.1109/TASE.2006.879686
11. Bassi, L., Secchi, C., Bonfé, M. and Fantuzzi, C. (2011) A SysML-Based Methodology for Manufacturing Machinery Modeling and Design. IEEE/ASME Transactions on Mechatronics, 16, 1049-1062. http://dx.doi.org/10.1109/TMECH.2010.2073480
12. Bonfé, M., Fantuzzi, C. and Secchi, C. (2013) Design Patterns for Model-Based Automation Software Design and Implementation. Control

Engineering Practice, 21, 1608-1619. http://dx.doi.org/10.1016/j.conengprac.2012.03.017

13. Thramboulidis, K. and Frey, G. (2011) Towards a Model-Driven IEC 61131-Based Development Process in Industrial Automation. Journal of Software Engineering and Applications, 4, 217-226. http://dx.doi.org/10.4236/jsea.2011.44024
14. Thramboulidis, K. (2012) IEC 61131 as Enabler of OO and MDD in Industrial Automation. IEEE International Conference on Industrial Informatics (INDIN), Beijing, 25-27 July 2012, 425-430.
15. Obermeier, M., Braun, S. and Vogel-Heuser, B. (2014) A Model Driven Approach on Object Oriented PLC Programming for Manufacturing Systems with Regard to Usability. IEEE Transactions on Industrial Informatics, 1.http://dx.doi.org/10.1109/TII.2014.2346133
16. Vogel-Heuser, B., Schütz, D., Timo, F. and Legat, C. (2014) Model-Driven Engineering of Manufacturing Automation Software Projects—A SysML-Based Approach. Mechatronics, 24, 883-897. http://dx.doi.org/10.1016/j.mechatronics.2014.05.003
17. Estévez, E. and Marcos, M. (2012) Model-Based Validation of Industrial Control Systems. IEEE Transactions on Industrial Informatics, 8, 302-310.http://dx.doi.org/10.1109/TII.2011.2174248
18. Estévez, E., Marcos, M., Iriondo, N. and Orive, D. (2007) Graphical Modeling of PLC-Based Industrial Control Applications. Proceedings of the 2007 American Control Conference, New York, 11-13 July 2007, 220-225.
19. Bartels, J. and Vogel, B. (2001) System Engineering Approach for Plant Automation (Systementwicklung für die Automatisierung im Anlagenbau). At-Automatisierungstechnik, 49, 214-224. http://dx.doi.org/10.1524/auto.2001.49.5.214
20. Land, M. and Horwood, J. (1995) Which Parts of the Road Guide Steering? Nature, 377, 339-340. http://dx.doi.org/10.1038/377339a0
21. Savioja, P. and Norros, L. (2013) Systems Usability Framework for Evaluating Tools in Safety-Critical Work. Cognition, Technology & Work, 15, 255-275.http://dx.doi.org/10.1007/s10111-012-0224-9
22. Siau, K. and Rossi, M. (2011) Evaluation Techniques for Systems Analysis and Design Modeling Methods—A Review and Comparative Analysis. Information Systems Journal, 21, 249-268. http://dx.doi.org/10.1111/j.1365-2575.2007.00255.x

23. Katzke, U., Vogel-Heuser, B. and Fischer, K. (2004) Analysis and State of the Art of Modules in Industrial Automation. ATP International-Automation Technology in Practice International, 46, 23-31.
24. Nielsen, J. (1993) Usability Engineering. Academic Press, Boston.
25. Obermeier, M., Schütz, D. and Vogel-Heuser, B. (2012) Evaluation of a Newly Developed Model-Driven PLC Programming Approach for Machine and Plant Automation. 8th IEEE International Conference on Systems, Man and Cybernetics (SMC), Seoul, 14-17 October 2012, 1552-1557.
26. Frank, U., Papenfort, J. and Schütz, D. (2011) Real-Time Capable Software Agents on IEC 61131 Systems-Developing a Tool Supported Method. Proceedings of 18th IFAC World Congress, Milan, 28 August-2 September 2011, 9164- 9169.
27. Strömman, M., Sierla, S. and Koskinen, K. (2005) Control Software Reuse Strategies with IEC 61499. 10th IEEE International Conference on Emerging Technologies & Factory Automation (ETFA), Catania, 19-22 September 2005, 749-756.
28. Sierla, S., Christensen, J., Koskinen, K. and Peltola, J. (2007) Educational Approaches for the Industrial Acceptance of IEC 61499. IEEE International Conference on Emerging Technologies & Factory Automation (ETFA), Patras, 25-28 September 2007, 482-489.
29. Patig, S. (2008) Preparing Meta-Analysis of Metamodel Understandability. Workshop on Empirical Studies of Model- Driven Engineering (ESMDE 2008), Toulouse, 29 September 2008, 11-20.
30. Patig, S. (2008) A Practical Guide to Testing the Understandability of Notations. Proceedings of the Fifth Asia-Pacific Conference on Conceptual Modelling, Wollongong, 79, 49-58.
31. Gemino, A. and Wand, Y. (2004) A Framework for Empirical Evaluation of Conceptual Modeling Techniques. Requirements Engineering, 9, 248-260.http://dx.doi.org/10.1007/s00766-004-0204-6
32. Vogel-Heuser, B. and Sommer, K. (2011) A Methodological Approach to Evaluate the Benefit and Usability of Different Modeling Notations for Automation Systems. Proceedings of the 7th IEEE International Conference on Automation Science and Engineering (CASE), Trieste, 24-27 August 2011, 474-481.
33. Lucas, M.R. and Tilbury, D.M. (2005) Methods of Measuring the Size and Complexity of PLC Programs in Different Logic Control Design Methodologies. The International Journal of Advanced Manufacturing Technology, 26, 436-447. http://dx.doi.org/10.1007/s00170-003-1996-0

34. Frey, G., Litz, L. and Klöckner, F. (2000) Complexity Metrics for Petri Net Based Logic Control Algorithms. IEEE International Conference on Systems, Man, and Cybernetics, Nashville, 2, 1204-1209.

35. Venkatesh, K., Zhou, M. and Caudill, R.J. (1994) Comparing Ladder Logic Diagrams and Petri Nets for Sequence Controller Design through a Discrete Manufacturing System. IEEE Transactions on Industrial Electronics, 41, 611- 619.http://dx.doi.org/10.1109/41.334578

36. Lee, J.S. and Hsu, P.L. (2001) A New Approach to Evaluate Ladder Diagrams and Petri Nets via the IF-THEN Transformation. IEEE International Conference on Systems, Man and Cybernetics, Tucson, 7-10 October 2001, 2711-2716.

37. Chidamber, S.R. and Kemerer, C.F. (1994) A Metrics Suite for Object Oriented Design. IEEE Transactions on Software Engineering, 20, 476-493.http://dx.doi.org/10.1109/32.295895

38. Michura, J. and Capretz, M.A.M. (2005) Metrics Suite for Class Complexity. IEEE International Conference on Information Technology: Coding and Computing (ITCC), Las Vegas, 4-6 April 2005, 404-409.

39. Fuchs, J., Feldmann, S., Legat, C. and Vogel-Heuser, B. (2014) Identification of Design Patterns for IEC 61131-3 in Machine and Plant Manufacturing. 19th IFAC World Congress, Cape Town, 24-29 August 2014, 6092-6097.

40. Park, E., Tilbury, D.M. and Khargonekar, P.P. (2001) A Modeling and Analysis Methodology for Modular Logic Controllers of Machining Systems Using Petri Net Formalism. IEEE Transactions on Systems, Man, and Cybernetics, Part C: Applications and Reviews, 31, 168-188. http://dx.doi.org/10.1109/5326.941841

41. Fuchs, J. (2011) Analyse und Neukonzeption der Softwaremodularität und deren Abbildung auf die maschinenbauliche Modularität am Beispiel eines Neuglasabschiebers der Lebensmittelindustrie. B.S. Thesis, Faculty of Mechanical engineering, TUM, Munich.

42. Recker, J.C., zur Muehlen, M., Keng, S., Erickson, J. and Indulska, M. (2009) Measuring Method Complexity: UML versus BPMN. Proceedings of the 15th Americas Conference on Information Systems, San Francisco, 6-9 August 2009, 1-10.

43. Rossi, M. and Brinkkemper, S. (1996) Complexity Metrics for Systems Development Methods and Techniques. Infor- mation Systems, 21, 209-227.http://dx.doi.org/10.1016/0306-4379(96)00012-9

44. Schalles, C. (2012) Usability Evaluation of Modeling Languages. Ph.D. Dissertation, Department of Computing, CIT, Cork.

45. Frank, T. (2014) Entwicklung und Evaluation einer Modellierungssprache für den Architekturentwurf von verteilten Automatisierungsanlagen auf Basis der Systems Modeling Language (SysML). Ph.D. Dissertation, Institute of Auto- mation and Information Systems, Technical University Munich, Munich.

46. Hajarnavis, V. and Young, K. (2008) An Assessment of PLC Software Structure Suitability for the Support of Flexible Manufacturing Processes. IEEE Transactions on Automation Science and Engineering, 5, 641-650. http://dx.doi.org/10.1109/TASE.2007.917135

47. Cross, J. and Denning, P. (2001) Computing Curriculum 2001. The Joint Curriculum Task Force IEEE-CS/ACM Report. http://www.acm.org/education/curric_vols/cc2001.pdf

48. Tucker, A., Deek, F., Jones, J., McCowan, D., Stephenson, C. and Verno, A. (2003) A Model Curriculum for K-12 Computer Science: Final Report of the ACM K-12 Task Force Curriculum Committee. The Association for Computing Machinery, New York.

49. Fitzgerald, S., McCauley, R., Hanks, B., Murphy, L., Simon, B. and Zander, C. (2010) Debugging from the Student Perspective. IEEE Transactions on Education, 53, 390-396.http://dx.doi.org/10.1109/TE.2009.2025266

50. Curtis, B. (1988) Five Paradigms in the Psychology of Programming. In: Helander, M., Ed., Handbook of Human- Computer Interaction, Elsevier North Holland, Amsterdam, 87-105. http://dx.doi.org/10.1016/B978-0-444-70536-5.50009-9

51. Magenheim, J., Nelles, W., Rhode, T., Schaper, N., Schubert, S. and Stechert, P. (2010) Competencies for Informatics Systems and Modeling: Results of Qualitative Content Analysis of Expert Interviews. IEEE Education Engineering, Madrid, 14-16 April 2010, 513-521.

52. Ruocco, A.S. (2003) Experiences in Threading UML throughout a Computer Science Program. IEEE Transactions on Education, 46, 226-228.http://dx.doi.org/10.1109/TE.2002.808263

53. Wickens, C.D. and Hollands, J.G. (2000) Engineering Psychology and Human Performance. 3rd Edition, Prentice Hall, Upper Saddle River.

54. Schweizer, K., Gramß, D., Mühlhausen, S. and Vogel-Heuser, B. (2009) Mental Models in Process Visualization— Could They Indicate the Effectiveness of an Operator's Training? Engineering Psychology and Cognitive Ergonomics, Springer, Berlin, Heidelberg, 297-306.

55. Gravetter, F.J. and Wallnau, L.B. (2006) Statistics for the Behavioral Sciences. Thomson/Wadsworth, Belmont.

56. Kim, J. and Lerch, F.J. (1992) Towards a Model of Cognitive Process in Logical Design: Comparing Object-Oriented and Traditional Functional Decomposition Software Methodologies. Proceedings of the SIGCHI Conference on Hu- man Factors in Computing Systems, Monterey, 3-7 July 1992, 489-498.

57. Kim, S.H. and Jeon, J.W. (2009) Introduction for Freshmen to Embedded Systems Using LEGO Mindstorms. IEEE Transactions on Education, 52, 99-108.http://dx.doi.org/10.1109/TE.2008.919809

58. Berges, M. and Hubwieser, P. (2011) Minimally Invasive Programming Courses: Learning OOP with (out) Instruction. Proceedings of the 42nd ACM Technical Symposium on Computer Science Education, Dallas, 9-12 March 2011, 7-92.

59. Faux, R. (2006) Impact of Preprogramming Course Curriculum on Learning in the First Programming Course. IEEE Transactions on Education, 49, 11-15.http://dx.doi.org/10.1109/TE.2005.852593

60. Jacobson, M.L., Said, R.A. and Rehman, H. (2006) Introducing Design Skills at the Freshman Level: Structured Design Experience. IEEE Transactions on Education, 49, 247-253. http://dx.doi.org/10.1109/TE.2006.872403

61. Verginis, I., Gogoulou, A., Gouli, E., Boubouka, M. and Grigoriadou, M. (2001) Enhancing Learning in Introductory Computer Science Courses through SCALE: An Empirical Study. IEEE Transactions on Education, 54, 1-13. http://dx.doi.org/10.1109/TE.2010.2040477

62. Lahtinen, E. (2007) A Categorization of Novice Programmers: A Cluster Analysis Study. Proceedings of the 19th An- nual Workshop of the Psychology of Programming Interest Group, Joensuu, 2-6 July 2007, 32-41.

63. Vogel-Heuser, B., Obermeier, M., Braun, S., Sommer, K., Jobst, F. and Schweizer, K. (2013) Evaluation of a UML- Based versus an IEC 61131-3-Based Software Engineering Approach for Teaching PLC Programming. IEEE Transac- tions on Education, 56, 329-335.http://dx.doi.org/10.1109/TE.2012.2226035

64. International Organization for Standardization (1999) Ergonomic Requirements for Office Work with Visual Display Terminals (VDTs)-Part 11: Guidance on Usability, EN ISO 9241-11:1998. Beuth, Berlin.

65. Bevan, N. (1995) Measuring Usability as Quality of Use. Software Quality Journal, 4, 115-130. http://dx.doi.org/10.1007/BF00402715

66. Annett, J. (2003) Hierarchical Task Analysis. In: Hollnagel, E., Ed., Handbook of Cognitive Task Design, Lawrence Erlbaum Assoc. Inc., Mahwah, 17-35.
67. Lucas, M.R. and Tilbury, D.M. (2002) Quantitative and Qualitative Comparisons of PLC Programs for a Small Testbed with a Focus on Human Issues. Proceedings of the American Control Conference, Anchorage, 5, 4165-4171.
68. Lucas, M.R. (2003) Understanding and Assessing Logic Control Design Methodologies. Ph.D. Dissertation, Dept. Mechanical Eng., University of Michigan, Ann Arbor.
69. Friedrich, D. and Vogel-Heuser, B. (2007) Benefit of System Modeling in Automation and Control Education. American Control Conference (ACC), New York, 9-13 July 2007, 2497-2502.
70. Friedrich, D. (2009) Anwendbarkeit von Methoden und Werkzeugen des konventionellen Softwareengineerings zur Modellierung und Programmierung von Steuerungssystemen. Ph.D. Dissertation, University Kassel, Kassel.
71. Friedrich, D. and Vogel-Heuser, B. (2005) Evaluating the Benefit of Modeling Notations on the Quality of PLC-Pro- gramming. 11th International Conference on Human-Computer Interaction (HCI), Las Vegas, 22-27 July 2005.
72. Katzke, U. and Vogel-Heuser, B. (2005) UML-PA as an Engineering Model for Distributed Process Automation. IFAC World Conference, Prague, 3-8 July 2005, 129-134.
73. Katzke, U. and Vogel-Heuser, B. (2009) Vergleich der Anwendbarkeit von UML und UML-PA in der anlagennahen Softwareentwicklung der Automatisierungstechnik-Beispiel einer vergleichenden empirischen Untersuchung von Mo- dellierungssprachen. at-Automatisierungstechnik, 57, 332-340. http://dx.doi.org/10.1524/auto.2009.0781
74. Witsch, D., Ricken, M., Kormann, B. and Vogel-Heuser, B. (2010) PLC-Statecharts: An Approach to Integrate UML- Statecharts in Open-Loop Control Engineering. 8th IEEE International Conference on Industrial Informatics (INDIN), Osaka, 13-16 July 2010, 915-920.
75. Witsch, D. and Vogel-Heuser, B. (2009) Close Integration between UML and IEC 61131-3: New Possibilities through Object-Oriented Extensions. IEEE International Conference on Emerging Technologies & Factory Automation (ETFA), Mallorca, 22-26 September 2009, 1-6.
76. Yang, S. and Sun, J.L. (2010) Modeling Traverse Feature in Concurrent Software System with UML Statecharts. IEEE International Conference

on Computational Intelligence and Software Engineering (CiSE), Wuhan, 10-12 December 2010, 133-138.

77. Witsch, D. (2012) Modellgetriebene Entwicklung von Steuerungssoftware auf Basis der UML unter Berücksichtigung der domänenspezifischen Anforderungen des Maschinen-und Anlagenbaus. Ph.D. Dissertation, Faculty of Mechanical Engineering, TUM, Munich.

78. Vogel-Heuser, B., Seidel, T., Braun, S., Obermeier, M., Sommer, K. and Johannes, C. (2011) Modeling Order Effects on Errors in Object Oriented Modeling for Machine and Plant Automation from an Educational Point of View. 16th IEEE International Conference on Emerging Technologies & Factory Automation (ETFA), Toulouse, 5-9 September 2011.

79. Frank, T., Eckert, K., Hadlich, T., Fay, A., Diederich, C. and Vogel-Heuser, B. (2012) Workflow and Decision Support for the Design of Distributed Automation Systems. IEEE INDIN: International Conference on Industrial Informatics, Beijing, 25-27 July 2012, 293-299. http://dx.doi.org/10.1109/INDIN.2012.6300859

80. Frank, T., Hadlich, T., Eckert, K., Fay, A., Diedrich, C. and Vogel-Heuser, B. (2012) Using Contact Points to Integrate Discipline Spanning Real-Time Requirements in Modeling Networked Automation Systems for Manufacturing Sys- tems. IEEE International Conference on Automation Science and Engineering (CASE), Seoul, 20-24 August 2012, 851-856.

81. Frank, T., Eckert, K., Hadlich, T., Fay, A., Diederich, C. and Vogel-Heuser, B. (2013) Erweiterung des V-Modells® für den Entwurf von verteilten Automatisierungssystemen. At-Automatisierungstechnik, 61, 79-91. http://dx.doi.org/10.1524/auto.2013.0009

82. Eckert, K., Hadlich, T., Fay, A., Diederich, C. and Vogel-Heuser, B. (2012) Design Patterns for Distributed Automa- tion Systems with Consideration of Non-Functional Requirements. 17th IEEE International Conference on Emerging Technologies & Factory Automation (ETFA), Krakow, 17-21 September 2012, 1-9.

83. Oswald, W.D. and Roth, E. (1978) Der Zahlen-Verbindungs-Test (ZVT). Ein sprachfreier Intelligenz-Schnell-Test. Verlag für Psychologie Horgrefe, Göttingen.

84. Hart, S. and Staveland, L. (1988) Development of NASA-TLX (Task Load Index): Results of Empirical and Theoretical Research. In: Hancock, P. and Meshkati, N., Eds., Human Mental Workload, North Holland, Amsterdam, 139-183.

85. Vogel-Heuser, B., Braun, S., Kormann, B. and Friedrich, D. (2011) Implementation and Evaluation of UML as Modeling Notation in Object

Oriented Software Engineering for Machine and Plant Automation. 18th World Congress of International Federation of Automation Control (IFAC), Milan, 28 August-2 September 2011, 9151-9157.

86. Ko, A.J. and Myers, B.A. (2005) A Framework and Methodology for Studying the Causes of Software Errors in Programming Systems. Journal of Visual Languages and Computing, 16, 41-84. http://dx.doi.org/10.1016/j.jvlc.2004.08.003

87. McCabe, T. (1976) A Complexity Measure. IEEE Transactions on Software Engineering, 2, 308-320. http://dx.doi.org/10.1109/TSE.1976.233837

Appendix A. Example of Subject's UML Model (Evaluation Scheme E1)

One example of a modular UML model (Figure A1) is given. The results show the subject's problem to identify reusable parts of the plant. He decided to use the state chart (Figure A1, left) and the class diagram (Figure A1, right). He tried to build a modular state chart and formed a class "single out and transport to stamp". This shows the inadequate understanding of classes and state charts and their application. Unfortunately, none of the subjects realized a correct class model.

Appendix B. (Evaluation Scheme E5 and WMC Calculation)

B.1. Evaluation Scheme E

In the following the measurement from E5 for plcUML and FBD model quality is shown. First the evaluated model elements are discussed for structure and behavior modeling. Then typical subject's solutions for both no- tations are shown and the points given are depicted.

The model quality regarding the grade of task completion (correct model elements) for both notations was evaluated for structure and behavior through manual code/model inspection. As measure for structure in plcUML models, the number of correct attributes with a correct data type and the correct access modifier in the class dia- gram was counted (cf. Figure B1). As methods were not imperative in order to solve the given task, they were not included in the measurement.

Additionally the created object instances in the main program were counted for the structure model. i.e. each correct instantiation of a cylinder object was counted (cf. Figure B2).

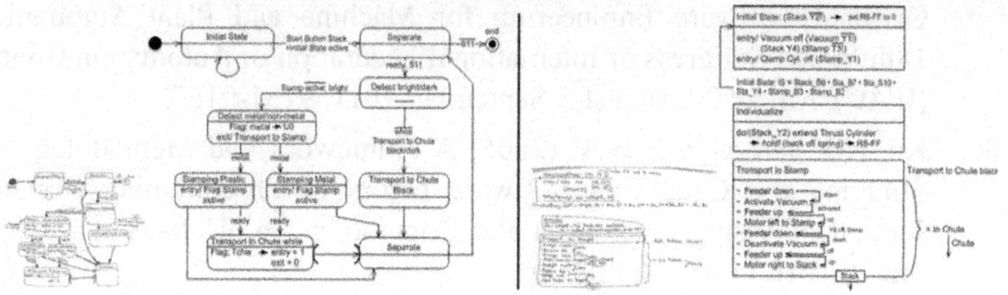

Figure A1: Modular UML behavior model of one subject (left: hand written model; right: translated model).

Cylinder	Attribute created	Correct datatype	Correct Access modifier
+ extended : BOOL	☑ 1P	☑ 1P	☑ 1P
+ RetExt_ref : BOOL	☑ 1P	☑ 1P	☑ 1P
+ retracted : BOOL	☑ 1P	☑ 1P	☑ 1P
+ capacitiveSensor_ Material_Detected : BOOL	☑ 1P	☑ 1P	☑ 1P
extend() retract()	4P	4P	4P max.: $\sum$ 12P

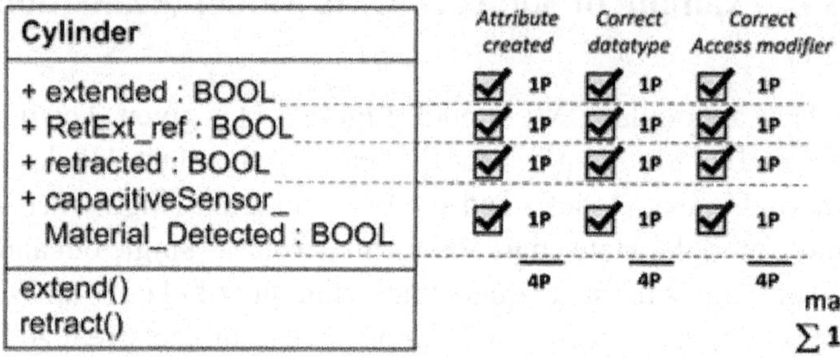

Figure B1: Structure model quality measurement plcUML: Class diagram.

		Instantiates 8 objects of class cylinder
1	**PROGRAM** MainProgram	
2	**VAR_INPUT**	
3	Cylinder1:Cylinder;	☑ 1P
4	Cylinder2:Cylinder;	☑ 1P
5	Cylinder3:Cylinder;	☑ 1P
6	Cylinder4:Cylinder;	☑ 1P
7	Cylinder5:Cylinder;	☑ 1P
8	StorageCylinder1:Cylinder;	☑ 1P
9	StorageCylinder2:Cylinder;	☑ 1P
10	StorageCylinder3:Cylinder;	☑ 1P
11	**END_VAR**	8P max.: $\sum$ 8P
12	**VAR_OUTPUT**	
13	**END_VAR**	

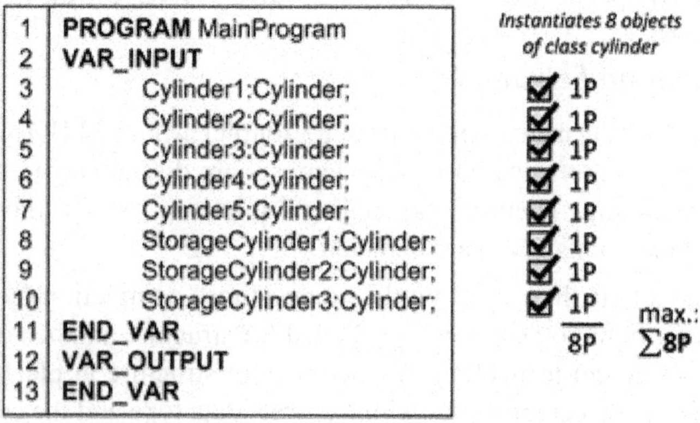

Figure B2: Structure model quality measurement plcUML: Object instantiation.

This results in a maximum of 20 points available for the structure model in plcUML. The plcUML behavior model quality was measured by identifying correct method calls, sequences of variable comparisons and states (cf. Figure B3). If the subsequent state after a logically correct variable comparison included a logically correct method call an additional point was given.

In Figure B4(a) and Figure B4(b) a complete example measurement for one student's model in UML is shown. Missing Points are depicted as Xs. The quality of the structure model (Figure B4(a)) is 13/20 or 65% and quality of the behavior model 24/67 or 35.82% (Figure B4(b)). The overall model quality is 37/87 or 42.53%.

Similar to the plcUML model quality measurement, the FBD program quality was evaluated. For the structure quality every necessary in- and output for the FBs was counted, cf. Figure B5. This results in a maximum of 32 points for FBD model structure quality.

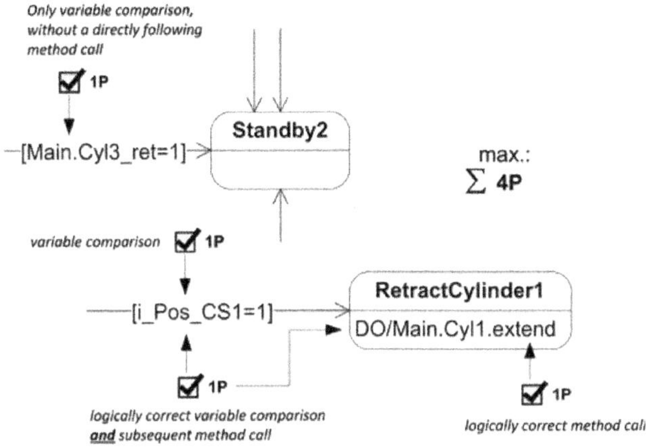

Figure B3: Behavior model quality measurement plcUML.

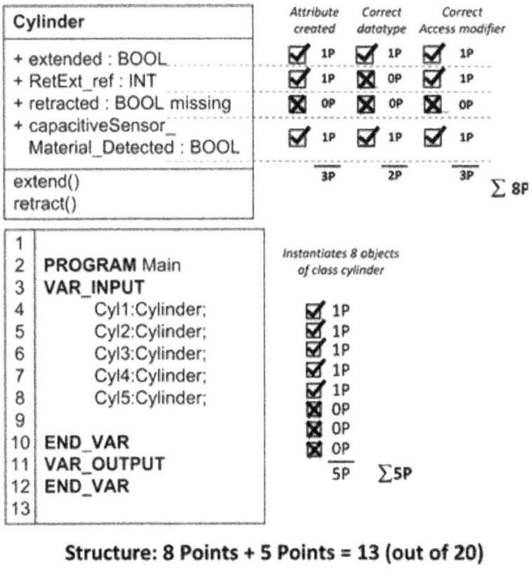

(a)

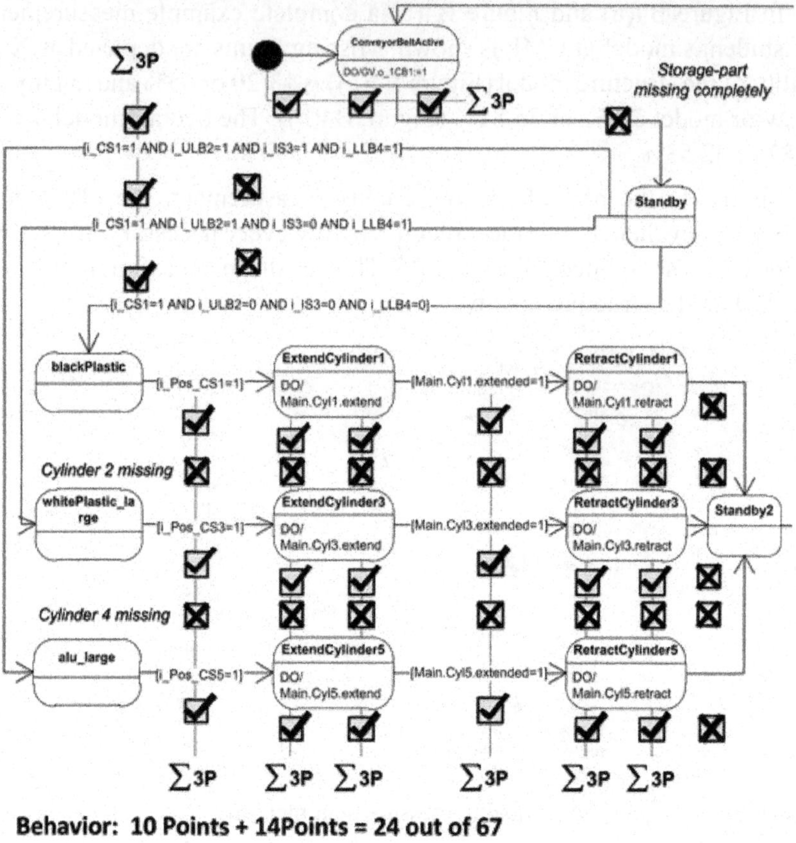

(b)

Figure B4. (a) plcUML structure model quality: 13/20 points; (b) plcUML behavior model example: model quality 47/67. Overall model quality: 13 + 24 = 37 Points/Relative overall modeling performance: 37/87 = 42.53%.

The FBD behavior model quality was measured by identifying correct FB or FC calls, sequences of variable comparisons and the connection of these elements, cf. Figure B6. If the subsequent call after a logically correct variable comparison included a logically correct FB or FC call an additional point was given.

B.2. WMC Calculation

WMC is defined as the sum of Ci. Ci is the cyclomatic complexity of the ith Method, and is calculated by counting the conditions of the method +1, cf. McCabe 1976 [87] . In Figure B7 an example for WMC calculation is given. In this case only the methods auto and manual of the example class are relevant. The auto method contains several conditions and therefore has a cyclomatic complexity corresponding to the number of included conditions +1 as defined by McCabe, resulting in Cauto = 10. The manual method does not contain any conditions resulting in a cyclomatic complexity Cmanual of 1. Finally these two complexity values sum up to an overall WMC of 11 for the example class.

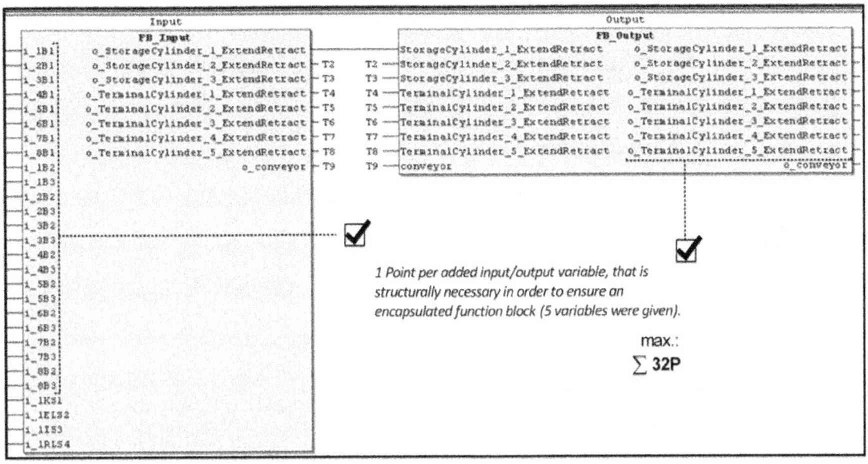

Figure B5: Structure model quality measurement FBD.

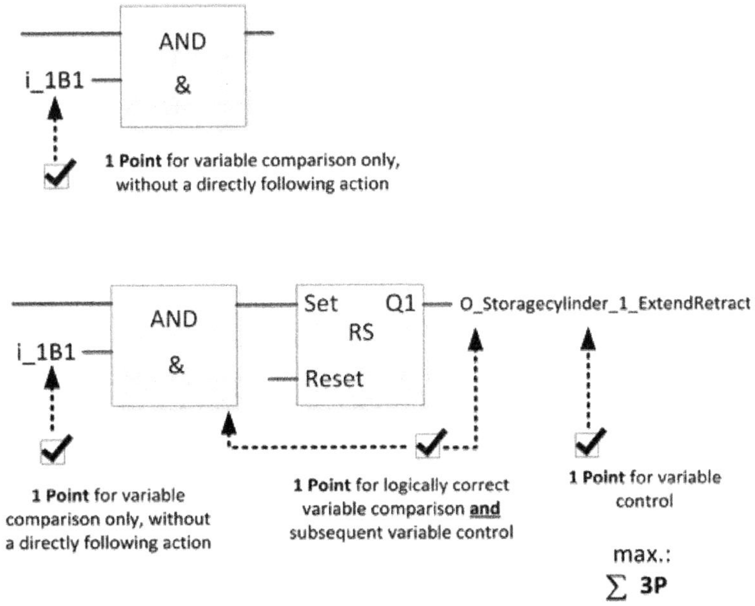

Figure B6: Behavior model quality measurement FBD.

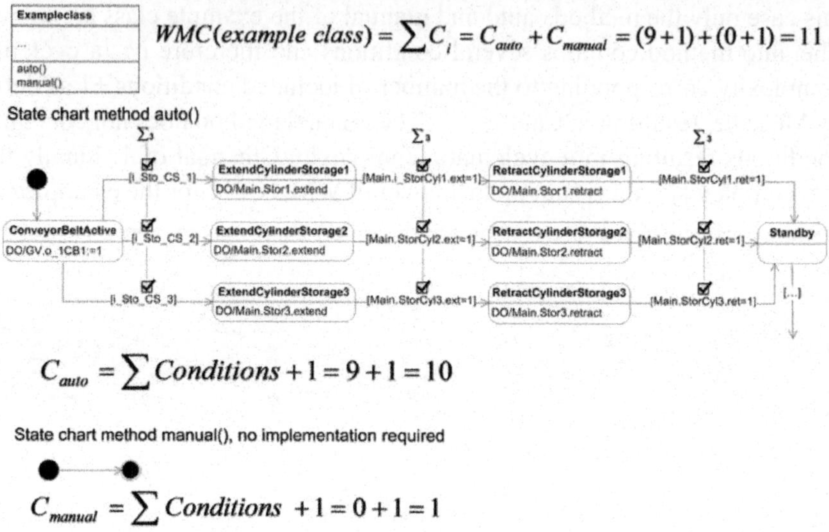

Figure B7: WMC calculation example.

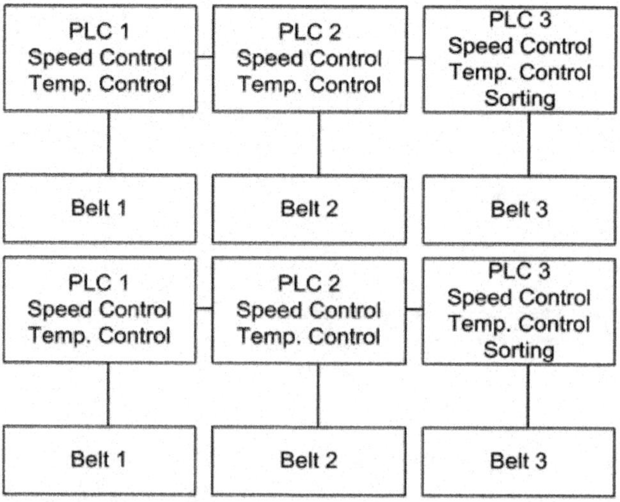

Figure C1: Most subject's functional oriented model (left) and Master model-mechatronic oriented model (right).

Appendix C. Master Model and Subjects' Solution (Evaluation Scheme E7)

Most subjects in E7 chose a functional oriented model deployment (Figure C1, left), i.e. deploying different functions (speed control, temperature control and sorting) on different PLCs, which is from an architectural pers- pective considering mechatronic modularity and reuse an inappropriate solution. NAS experts chose a different approach (Figure C1, right) to support reuse of existing modules. A comparison and evaluation of these differ- ent models was difficult as "how to build a module" was not included in the training, because the MDE ap- proach should be intuitively applicable.

The best subject gained 144 out of 182 points building a mechatronic oriented model, but neglected requirements and other details.

Chapter 5

A CRITICAL REVIEW OF MACHINE LOADING PROBLEM IN FLEXIBLE MANUFACTURING SYSTEM

Ranbir Singh[1], Rajender Singh[2], B. K. Khan[3]

[1]Research Scholar, Department of Mechanical Engineering, DCRUST Murthal, Sonepat, India

[2]Professor, Department of Mechanical Engineering, DCRUST Murthal, Sonepat, India

[3]Technical Advisor, MSIT, Sonepat, India

ABSTRACT

Production planning is the foremost task for manufacturing firms to deal with, especially adopting Flexible Manufacturing System (FMS) as the manufacturing strategy for production seeking an optimal balance between productivity-flexibility requirements. Production planning in FMS provides a solution to problems regarding part type selection: machine grouping, production ratio, resource allocation and loading problem. These problems need to be solved optimally for maximum utilization of resources. Optimal solution to these problems has been a focus of attention in production and manufacturing, industrial and academic research since a number of decades. Evolution of new optimization techniques, software, technology, machines and computer languages provides the scope of a better optimal solution to the existing problems. Thus there remains a need of research to solve the problem with latest tools and techniques for higher optimal use of available resources. As an objective, the researchers need to reduce the computational time and cost, complexity of the problem, solution approach viz. general or customized, better user friendly communication with machine, higher freedom to select the desired objective(s) type(s) for optimal solution to the problem. As an approach to the solution to the problem, researcher first needs to go for an exhaustive literature review, where the researcher needs to find the research gaps, compare and analyze the tools and techniques used, number of objectives considered for optimization and need, and scope of research for the research problem. The present study is a review paper analyzing the research gaps, approach and techniques used, scope of new optimization techniques or any other research, objectives considered and validation approaches for loading problems of production planning in FMS.

INTRODUCTION

Manufacturing is the pilot element within the overall enterprise. Possible manufacturing outputs of the firm to meet pre-determined corporate level goals should be known to remain in competition at global market. Manufacturing strategy writes the script to calculate possible manufacturing outputs. Existence of the manufacturing strategy guides daily decisions and activities with clear understanding of decision-goal relationship of the corporation and provides a vision for the firm to remain aligned with the overall business strategy of the firm. The firms having manufacturing strategies for achieving corporate goals survive for long run. A strategy is also a strong communication tool between different levels of management to bring all operations in line with corporate objectives. Custom manufacturing, continuous manufacturing, intermittent manufacturing, flexible manufacturing, just-in-time manufacturing, lean manufacturing and agile manufacturing are major manufacturing strategies revealed in the literature.

FMS is an automated manufacturing system consisting of computer numerical control (CNC) machines with automated material handling, storage and retrieval system. The aim of FMS is to attain the efficiency of mass production while utilizing the flexibility of job shop simultaneously. FMS is adopted for batch production of mid production volume and mid part variety (flexibility) requirements. Since its evolution, researchers are working for optimality of FMS strategy. FMS is a field of great potential hence a numerous complex planning problems need to be solved. Major complex production planning problems are part type selection: machine grouping, production ratio, resource allocation and loading problem (Stecke, 1983). All the production planning problems need to be optimally solved. The present research is the critical literature review for the loading problem of production planning in FMS.

Tooling individual or group of machine(s) to collectively accomplish all manufacturing operations concurrently for all part type in a batch is termed as loading problem. A solution to the problem specifies the machine(s) to which a job has to be routed in sequence for each of its operation(s) with respective tooling under capacity and technological constraint(s) for all jobs in a batch simultaneously to achieve certain objective(s). Loading is a complex combinational planning problem because a batch of jobs is to be machined simultaneously and each job requires unique set of operations effect on manufacturing cost.

To solve the problem, highly experienced and skilled professionals are required. Without the use of some computational or optimization technique, the solution may or may not be optimal. Thus there arises the need of optimal solution with the help of computational methods using optimization techniques. The paper is a critical review paper analyzing the research gaps, approach and techniques used, scope of new optimization techniques or any other research, objectives considered and validation approaches for loading problems of production planning in FMS.

LITERATURE REVIEW OF LOADING PROBLEMS IN FMS

In brief, to solve a problem using optimization techniques and computational analysis, objective(s) are first set, the physical system is modelled using certain technique like mathematical modelling, the solution is then derived under given boundary conditions and constraints to achieve the given objectives, the results are then analysed and the solution approach is then validated. Heuristics has been widely used by the researchers. Table 1 presents the tabulated research review discussing the approach, objectives and results of the loading problems in FMS. Flexible manufacturing is an overall pilot element within an enterprise. Each multinational manufacturing concern has to satisfy business goals to remain in competition with the global market. The manufacturing firm should be aware of the possible manufacturing outputs that will closely match the goals and strategy determined at the corporate level. The existence of a manufacturing strategies guide the daily decisions and activities with clear understanding of how those daily decisions relate to the overall goals of the corporation. The firms having manufacturing strategies for achieving corporate goals survive long. A manufacturing strategy provides a vision to the manufacturing organization for keeping itself aligned with the overall business strategy of the corporation. A strategy is also a strong communication tool between different levels of management to bring all operations in line with corporate objectives. Custom manufacturing, continuous manufacturing, intermittent manufacturing, flexible manufacturing, just-in-time manufacturing, lean manufacturing and agile manufacturing are the major manufacturing strategies which are revealed in literature.

FMS is an automated manufacturing system consisting of numerical control (computer) machines with automated material handling, automated storage and retrieval system. The aim of FMS is to attain the efficiency of mass production while utilizing the flexibility of job shop simultaneously.

Most of the researches are focused on increasing the production volume of FMS with increased part varieties. FMS is an interesting field of research to solve the issues and problems encountered by industries. Though FMS has great potential benefits, a numerous control and planning problem need to be taken care of. Kathryn E. Stecke in 1983 described five complex productions planning problems namely part type selection problem, Machine Grouping Problem, production ratio problem, resource allocation problem and loading problem [1] .

Table 1: Review of machine loading problems in FMS based on heuristics approach

HEURISTIC APPROACH						
Sr.	Year	Researcher Name	Approach	Objectives	Results	Validation approach
1	1983	K. E. Stecke & F. Brian Talbot [2]	Heuristic methods	Ø Minimizing part movements Ø Balancing of workload Ø Unbalancing of workload	Determined how machine tool magazine in a FMS can be loaded to meet simultaneous requirements of a number of different parts	Computational results are presented
2	1985	K. Shankar & Y. J. Tzen [14]	Heuristic methods	Ø Minimizing system unbalance Ø Number of late jobs Ø Balancing of workload	Computational results presented gives improved results	performance is compared with previous results from literature
3	1988	J. A. Ventura, F. F. Chen, & M. S. Leonard [15]	Heuristic algorithms	Ø Minimizing make span	Improved Performance	The performance of each of the proposed algorithms is evaluated by testing on two hypothetical FMSs.
4	1990	B. Ram, S. Sarin, & C. S. Chen [16]	Fast heuristic algorithms	Ø Maximizing throughput	FMS loading problem can be solved near optimally in short time	Computational results are produced and compared with previous results
5	1992	S. K. Mukhopadhyay, S. Midha, & V. Murlikrishna [17]	Heuristic procedure	Ø Minimizing system unbalance Ø Maximizing throughput	Results show that algo developed is very reliable and efficient	tested on ten problems and are compared with existing results
6	1993	K. Kato, F. Oba, & F. Hashimoto [18]	Heuristic approach	Ø Minimizing total number of cutting tools required Ø Maximizing utilization rate of each machine	Computational results shows improved effectiveness	Computational results are given to demonstrate the effectiveness of the proposed method
7	1995	E. K. Steeke & F. Brian Talbot [19]	Heuristic Algorithms	Ø Minimizing part movements Ø balancing of workload Ø unbalancing of workload	Results are computationally demonstrated & found improved significantly	Computational results are produced and compared with previous results
8	1997	M. K. Tiwari et al. [20]	Heuristic solution approach	Ø Maximizing throughput Ø Minimizing system unbalance	Graphical representation and subsequent model validation	Computational results are produced and compared with previous results
9	1998	G. K. Nayak & D. Acharya [21]	Heuristics and mathematical programming approaches	Ø Maximizing part types in each batch Ø Maximizing routing flexibility of batches	Heuristic proposed for part type selection & simple mathematical programs for other two problems	Computational results are compared with existing results
10	2000	D.-H. Lee & Y.-D. Kim [22]	Heuristic algorithms	Ø Minimizing maximum workload of machines	Results show that suggested algos perform better than existing	Simulation results are compared with existing results

11	2006	N. Nagarjunaa, O. Maheshb, & K. Rajagopal [23]	Heuristic based on multi stage programming approach	Ø Minimizing system unbalance	Bring together productivity of flow lines and flexibility of job shops	Tested on 10 sample problems available in FMS literature and compared with existing solution methods
12	2006	M. Goswami & M. K. Tiwari [24]	Heuristic-based approach	Ø Minimizing system unbalance Ø Maximizing throughput	Loading problem is crucial link between tactical planning and operational decisions	Extensive computational experiments have been carried out to assess the performance of the proposed heuristic and validate its relevance
13	2007	M. K. Tiwari, J. Saha, & S. K. Mukhopadhyay [25]	Heuristic Solution Approaches	Ø Minimizing system unbalance Ø Maximizing throughput	GA based heuristic are found more efficient and outperform in terms of solution quality	tested on problems representing three different FMS scenarios from available literature

Loading means allocation of the operations and required tools to a part types among the set of machine(s), subjected to resource & technological constraints to collectively accomplish all manufacturing operations for each pat type machined concurrently. The allocation of workloads to the existing production facilities for manufacturing products with several constraints in order to perform production activities according to the production plan established, it is essential to adjust the workload for each of the facilities and workers in each time period so they are not assigned work exceeding the given capacity. A solution to this problem specifies the tools which must be loaded in each machine tool magazine and the machine(s) to which a part can be routed for each of its operations before production begins. A variety of products are manufactured simultaneously in FMS, where each part requires potentially unique set of operations, and loading problem is declared as a combinational problem by Kathryn E. Stecke [2] which is highly complex, time-consuming and tedious in nature & requires highly experienced process planners.

Machine loading is one of the most critical production planning problems of FMS. It concerns with the time spend by the job(s) on machine(s) and the manufacturing cost. Manufacturing cost is the sum of fixed and variable costs. Variable cost varies with the level of production output. As output increases, variable cost increases. Once invested, we can't play around the fixed cost; hence to reduce the manufacturing cost, researcher has to minimize the variable cost while maximizing the output. This is done by developing and optimizing a virtual model of manufacturing by some conventional or non-conventional technique for certain number of objectives with their individual weightage accordingly. A researcher has to solve the manufacturing model to minimize the time spent by the job on machine, number of tool used, and movements of tool and job. FMS is a group technology concept hence all the operations on the group jobs are required to be completed at once keeping in view that no machine should be idle or overloaded at any instance of time. Thus the optimized solution of the machine loading problem for certain objectives under technological and capacity constraints is required. The solution to the machine loading problem is to minimize the manufacturing cost as a whole.

Increasing part varieties with raised productivity is necessary to be in competition and to maintain the demand of the product, which is possible by continuous research and optimized solutions to each of the production planning problem. This paper presents a research review of the optimization techniques and the objectives for which the machine loading problem in FMS has been solved, and scope of the research in the field.

Before presenting literature review, an introduction to the optimization techniques and their classification seems necessary to be discussed here for better understanding of the subject. Optimization is the approach for ideal solution. Accuracy of the solution depends on the approach, modeling, computational time and capacity, and nature of the problem. Optimization is classified into six categories: function, and trial and error, single variable and multiple variables, static and dynamics, continuous and discrete, constrained and unconstrained, and random and minimum seeking.

A functional optimization is for theoretical approach where a mathematical formula describes the objective function. Trial-and-error optimization is for experimental optimization with change in the variables which affect output without knowing much about the process. An optimization can be single variable for one dimensional analysis, and multi variables for multi-dimensional analysis. As the number of variables increases, the complexity of the problem also increases. Static optimization is independent of time and dynamic optimization as a function of time. Discrete optimization has a finite number of variables with all possible values, while continuous optimization has infinite number of variables with all possible values. Values are incorporated in equalities and inequalities to an objective of variable function in constrained optimization while the variables can take any value in unconstrained optimization. Random optimization finds sets of variables by probabilistic calculations while minimal seeking is the traditional optimization algorithms which are generally based on calculus methods and minimizes the function by starting from an initial set of variable values.

These optimization approaches can be further sub-categorized as stochastic programming, integer programming, linear programming, nonlinear programming, bound programming, network programming, least squares methods, global optimization, and non-differential optimization.

Most of the researches are focused on solving the machine loading problem by global optimization algorithms. Global optimization algorithms are generally categorized into two approaches: deterministic and probabilistic. Deterministic are sub-categorized into static space search (1992) [3] , branch and bond and algebraic geometry algorithms. Probabilistic is sub-categorized as Monte Carlo algorithms, soft computing and Artificial Intelligence (AI). Monte Carlo algorithms includes two classes, one covers Stochastic (hill climbing) (2002) [4] , Random optimization (1963) [5] , Simulated Annealing (SA) (1953) [6] , Tabu Search (TS) (1989) [7] , Parallel tempering, Stochastic tunneling and Direct Monte Carlo Sampling, and second class includes Evolutionary Computation (EC). EC can be performed by Monte Carlo algorithms or soft computing or AI. EC is further classified as Evolutionary Algorithms (EA), Memetic Algorithms (hybrid Algorithms) (1989) [8] , Harmonic Search (HS), Swarm Intelligence (SI). EA is sub-classified as Genetic Algorithms (GA) (1962) [9] , Learning Classifier System (LCS) (1977) [10] , Evolutionary Programming, Evolution Strategy (ES), Genetic Programming (GP) (1958) [11] . ES includes Differential Evolution (DE), and GP includes Standard GP, Linear GP and Grammar Guided GP. SI includes Ant Colony Optimization (ACO) (1996) [12] and Particle Swarm Optimization (PSO) (1995) [13] . The above discussed classifications scheme will be used for classifying the optimization techniques for solving the machine loading problems of FMS in the paper. Figure 1 shows the evolution of the major optimization techniques along the time axis.

LITERATURE REVIEW OF MACHINE LOADING PROBLEMS IN FMS

An exhaustive research review has been carried out for study of approaches and optimization techniques for machine loading problems in FMS. A. Baveja, A. Jain, A. K. Singh, A. Kumar, A. M. Abazari, A. Murthy, A. Prakash, A. Srinivasulu, A. Turkcan, C. A. Yano, C. Basnet , C.S. Chen, D. Acharya, D. Kosucuoglu, D.H. Lee, F. Brian Talbot, F. F. Chen, F. Guerrero, F. Hashimoto, F. Oba, G. K. Nayak, G.C. Lee, H. C. Co, H. Sattari, H. Yong, H.B. Jun, H.-K. Roh, J. A. Ventura, J. Larranaeta, J. S. Biermann, J. Saha, J. G. Shanthikumar, J. N. D. Gupta, K Chandrashekara, K. E. Stecke, K. Kato, K. M. Bretthauer, K. Rajagopal, K. Shankar, L. H. S. Luong, L. S. Kiat, M. A. Gamila, M. A. Venkataramanan, M. Arıkan, M. Berrada, M. Goswami, M. I. Mgwatua, M. K. Pandey, M. K. Tiwari, Ming Liang, M. M. Aldaihani, M. S. Akturk, M. S. Leonard, M. Savsar, M. Solimanpur, M. Yogeswaran, N. K. Vidyarthi, N. Khilwani, N. Kumar, N. Nagarjunaa, N. K. Vidyarthi, O. Maheshb, Prakash, R. P. Sadowski, R. Budiarto, R. D. Matta, R. H. Storer, R. M. Marian, R. R. Kumar, R. Shankar, R. Swarnkar, S. Biswas, S. Deris, S. Erol, S. G. Ponnambalam, S. K. Mandal, S. K. Mukhopadhyay, S. Kumar, S. Lozano, S. Midha, S. Motavalli,

S. P. Dutt, S. Rahimifard, S. S. Mahapatra, S.C. Sarin, S.K. Chen, S.K. Lim, S.T. Newman, T. J. Greene, T. J. Sawik, T. Koltai, T. L. Morin, T. Sawik, U. Bilge, U. K. Yusof, V. H. Nguyen, V. M Kumar, V. Murlikrishna, V. N. Hsu, V. Tyagi, W. F. Mahmudy, Y. Cohen, Y. D. Kim, Y. J. Tzen and Z. Wu are key researchers for solving the loading problem of production planning in FMS.

The tabulated research review discussing the approach, objectives, results and validation approach for machine loading problems in FMS is discussed in Tables 1-3. The literature review is classified into three groups: (1) heuristics; (2) global optimization; and (3) other optimization techniques.

Table 1 presents the review of machine loading problems of FMS based on heuristics approach. The heuristics approach has been significantly used for solving the research problem. Research has gained significant acceleration with the evolution and growth of global optimization techniques.

Table 2 presents the review of machine loading problems in FMS based on global optimization algorithms. Global optimization techniques have been explored rigorously by the researchers. The natural selection techniques have reported good results compared to others. The application of global optimization techniques for solving machine loading problem is increasing with growth of natural optimization techniques. The results reported by natural optimization techniques are more acceptable. Natural optimization techniques, GA and PSO are widely used techniques.

Table 3 presents the review of machine loading problems in FMS based on optimization techniques not falling in the above classification. Since the major focus is on heuristics and global optimization techniques, thus other techniques are grouped in a single table. These techniques have been adopted from time to time for solving the machine loading problem as shown year wise in Table 3.

Optimization techniques and approaches under the classification of global optimization scheme are discussed in Table 2.

Optimization techniques and approaches not falling under the above classifications are discussed inTable 3.

Table 4 has been formulated on regressive analysis of Tables 1-3, for the analysis of the loading objectives to be fulfilled while solving the loading problem. It is a year-wise tabulation and analysis of the loading objectives.

Table 2: Review of machine loading problems in FMS based on global optimization algorithms

GLOBAL OPTIMIZATION ALGORITHMS						
a. Deterministic approach						
1. Branch and bound						
Sr.	Year	Researcher Name	Approach	Objectives	Results	Validation approach
1	1986	M. Berrada & K. E. Stecke [26]	Branch and bound approach	Ø Balancing of workload	Computational results gives fruitful results	Computational results are produced and demonstrated the efficiency of suggested procedures
2	1989	K. Shankar & A. Srinivasulu [27]	Branch & backtrack procedure and Heuristic procedures	Ø Maximizing assigned workload Ø Maximizing throughput Ø Minimizing workload unbalance	Each procedure is illustrative by numerical example and results are with improved performance	An illustrative numerical example
3	1994	Y. D. Kim & C. A. Yano [28]	New branch and bond algorithm	Ø Maximizing throughput	Improved efficiency	Computational results are produced and compared with previous results
2. Algebraic Geometry						
4	1986	T. J. Greene & R. P. Sadowski [29]	Mixed integer programming	Ø Minimizing make span Ø Minimizing mean flow time Ø Minimizing mean lateness	Explained simple numeric example	a simple numeric example
5	1987	S.C. Sarin & C.S. Chen [30]	Mathematical model	Ø Minimizing overall machining cost	Computational results are reported	Computational results are compared with literature results
6	1990	K. M. Bretthauer & M. A. Venkataramanan [31]	Linear Integer Programming	Ø Maximizing weighted sum of number of operation to machine assignments	Computational results are satisfactory with improved performance	Computational results are produced
7	1990	H. C. Co, J. S. Biermann, & S.K. Chen [32]	Mixed-integer programming (MIP)	Ø Balancing of workloads	Results were found practical	Computational results are produced
8	1990	M. Liang & S. P. Dutt [33]	Mixed-Integer Programming	Ø Minimizing production cost	Demand for change on optimal solution	An example problem is solved
9	1993	Ming Liang [34]	Non-linear programming	Ø Maximizing system output	production cost can be significantly reduced using this approach	Computational results with an illustrative example is demonstrated
10	1994	Ming Liang [35]	Non-linear programming	Ø Maximizing system output Ø minimizing production cost	Production cost can be significantly reduced using this approach	An illustrative example is solved using the suggested approach
11	1997	V. N. Hsu & R. D. Matta [36]	Lagrangian-based heuristic procedure (MIP problem formulation)	Ø total processing cost	finds a good loading solution	iteratively compared different scenarios
12	1998	T. J. Sawik [37]	Integer programming & approximative lexicographic approach	Ø Balancing workloads Ø Minimizing total interstation transfer time	Results of computational experiments are reported	illustrative example and some results of computational experiments
13	1999	F. Guerrero, S. Lozano, T. Koltai, & J. Larranaeta [38]	Mixed-integer linear program	Ø Balancing of workload	New approach to loading problem	Computational results are produced
14	2001	N. Kumar & K. Shanker [39]	Mixed integer programming	Ø Balancing of Workload	Results are in agreement with previous findings	Computational results are compared with the previous findings

15	2003	M. A. Gamila & S. Motavalli [40]	mixed integer programming	Ø Minimizing completion time Ø Minimizing Material handling time Ø Minimizing total processing time	Results reported increased efficiency and performance of system	Computational results are compared with the previous findings
16	2004	T. Sawik [41]	Mixed integer programming	Ø Minimizing production time	Computational results reported better performance	Numerical examples and some computational results are compared with available literature
17	2011	M. I. Mgwatua [42]	Linear Mathematical Programming	Ø Maximizing throughput Ø Minimizing make span	More interactive decisions and well-balanced workload of the FMS can be achieved when sub-problems are solved jointly	Compared with results from previous literature
18	2012	A. M. Abazari, M. Solimanpur, & H. Sattari [43]	Linear mathematical programming	Ø Minimizing System unbalance	Genetic algorithm (GA) is proposed and performance of proposed GA is evaluated based on some benchmark problems	Performance is evaluated based on some benchmark problems adopted from the literature
				b. Probabilistic		
				3. Monte Carlo algorithms		
19	1998	S. K. Mukhopadhyay et al. [44]	Simulated annealing (SA) approach	Ø Minimizing system imbalance	Tried to give global optimum solution	Computational results are compared with existing results
20	2004	R. Swarnkar & M. K. Tiwari [45]	Hybrid tabu search and simulated annealing based heuristic approach	Ø Minimizing system unbalance Ø Maximizing throughput	Results reported better performance	Tested on Standard problems and the results obtained are compared with those from some of the existing heuristics from literature
21	2005	M. M. Aldaihani & M. Savsar [46]	Stochastic model	Ø Minimizing total (FMC) flexible manufacturing cell cost per unit of production	Results reported better performance	Computational results were presented
22	2006	M. K. Tiwari, S. Kumar, S. Kumar, Prakash, & R. Shankar [47]	Constraints-Based Fast Simulated Annealing (SA) Algorithm	Ø Minimizing system unbalance Ø Maximizing throughput	Proposed algorithm enjoys the merits of simple SA and simple genetic algorithm	The application of the algorithm is tested on standard data sets
23	2012	M. Arıkan & S. Erol [48]	Hybrid simulated annealing-tabu search algorithm	Ø Maximizing weighted sum Ø Minimizing system unbalance Ø Balancing of workload	Results shows improved system performance compared to earlier results in literature	The results are compared with those developed earlier by the authors
				4. Evolutionary Computation (EC)		
				ü Evolutionary algorithms (EA)		
24	2000	N. Kumar & K. Shanker [49]	Genetic algorithm (GA)	Ø Maximizing number of part types in a batch Ø Maximizing number of parts selected a batch Ø Maximizing mean machine utilization	Results reported reduced computational requirements	comparative study of Computational results
25	2002	H. Yong & Z. Wu [50]	GA-based integrated approach	Ø Balancing of workloads	Results shows that suggested approach perform better than existing	Computational results are compared with the previous findings

26	2006	A. Kumar, Prakash, M. K. Tiwari, R. Shankar, & A. Baveja [51]	Constraint based genetic algorithm (CBGA)	Ø Balancing machine processing time Ø Minimizing number of movements Ø Balancing of workload Ø Unbalancing of workload Ø Filling the tool magazines as densely as possible Ø Maximizing sum of operations priorities	The methodology developed here helps avoid getting trapped at local minima	The application of the algorithm is tested on standard data sets from available literature.
27	2007	A. Turkcan, M. S. Akturk, & R. H. Storer [52]	Genetic Algorithm (GA)	Ø Minimizing manufacturing cost Ø Total weighted tardiness	Approach improves CNC machine efficiency & responsiveness to customer due date requirements	compared with the performance of most commonly used approach in the literature
28	2008	V. Tyagi & A. Jain [53]	Genetic algorithm based methodology	Ø Minimizing system unbalance	For a given number of tool copies of each tool type tool loading is affected by the availability of flexible process plans	An illustrative example
29	2012	U. K. Yusof, R. Budiarto, & S. Deris [54]	Constraint-chromosome genetic algorithm	Ø Minimizing system unbalance Ø Maximizing throughput	Overall combined objective function increased by 3.60% from previous best result	tested on 10 sample problems available in the FMS literature and compared with existing solution methods
			ü Memetic (hybrid) Algorithms			
30	2000	M. K. Tiwari & N. K. Vidyarthi [55]	Genetic Algorithm (GA) based (HA) Heuristic Approach	Ø Minimizing system unbalance Ø Maximizing throughput	Optimal solution to problem	Tested on ten sample problems and the computational results obtained have been compared with those of existing methods
31	2009	M. Yogeswaran, S. G. Ponnambalam, & M. K. Tiwari [56]	Hybrid genetic algorithm simulated annealing algorithm (GASAA)	Ø Minimising system unbalance Ø Maximising throughput	Results support better performance of GASA over algorithms reported in literature	results compared with reported in the literature
32	2010	S. K. Mandal, M. K. Pandey, & M. K. Tiwari [57]	Genetic algorithm simulated annealing Heuristics approach	Ø Minimizing breakdowns Ø Minimizing system unbalance Ø Minimizing make span Ø Maximizing throughput	Results incurred under breakdowns validate robustness of developed model for dynamic ambient of FMS	Compared with dataset from previous literature
33	2012	V. M Kumar, A. Murthy, & K Chandrashekara [58]	Meta-hybrid heuristic technique based on genetic algorithm and particle swarm optimization	Ø Minimizing system unbalance Ø Maximizing throughput	Model efficiency and performance of system is comparable with results compared to literature	Computational results are presented
34	2012	C. Basnet [59]	Hybrid genetic algorithm	Ø Minimizing system unbalance	Better solutions for system unbalance	Computational comparison between the genetic algorithm and previous algorithms is presented
35	2012	D. Kosucuoglu & U. Bilge [60]	Genetic algorithm based mathematical programming (GAMP)	Ø Minimizing total distance travelled by parts during production	GALP integration works successfully for this hard-to-solve problem	tested through extensive numerical experiments
				ü Swarm Optimization		
36	2007	S. Biswas & S. S. Mahapatra [61]	Swarm Optimization Approach	Ø Minimizing system unbalance	Results reported improved system balance	compared with existing techniques for ten standard problems available in literature representing three different FMS scenarios

37	2008	S. Biswas & S. S. Mahapatra [62]	Modified particle swarm optimization	Ø Minimizing system unbalance	Proposed algorithm produces promising results in comparison to existing methods	comparison to existing methods for ten benchmark instances available in the FMS literature
38	2008	S. G. Ponnambalam & L. S. Kiat [63]	Particle Swarm Optimization (PSO)	Ø Minimizing system unbalance Ø Maximizing throughput	Performance of PSO is satisfactory compared with heuristics reported in literature	tested by using 10 sample dataset and the results are compared with the heuristics reported in the literature
				ü Artificial intelligence		
39	2001	N. K. Vidyarthi & M. K. Tiwari [64]	Fuzzy-based Heuristic Approach	Ø Minimizing system unbalance Ø Maximizing throughput	Substantial improvement in solution quality over some existing heuristic-based approaches	Tested on 10 problems adopted from literatures and computational results are compared with the previous findings
40	2004	R. R. Kumar, A. K. Singh, & M.K. Tiwari [65]	Fuzzy based algorithm	Ø Minimizing system unbalance Ø Maximizing throughput	Extended neuro fuzzy petri net is constructed	Computational results are compared with standard data set adopted from literature
41	2008	A. Prakash, N. Khilwani, M. K. Tiwari, & Y. Cohen [66]	Modified immune algorithm	Ø Maximizing throughput Ø Minimizing system unbalance	Good results as compared to best results reported in literature	compared to the best results reported in the literature

The table is showing the list of objectives for which the loading problem is solved. The tick mark (√) in the table shows the density for repeatability of the objectives.

Abbreviations used in Table 4:

1) Minimizing system unbalance
2) Maximizing throughput
3) Balancing of workload in the system configured of groups composed of machines of equal size
4) Minimizing make span
5) Meeting delivery dates
6) Minimizing manufacturing cost/Minimizing total processing cost/ Minimizing total flexible manufacturing cell cost per unit of production
7) Minimizing tardiness
8) Minimizing production cost
9) Unbalancing the workload per machine for a system of groups of pooled machines of unequal sizes
10) Minimizing part movements
11) Maximizing part types in each batch
12) Minimizing subcontracting costs
13) Maximizing weighted sum of number of operation to machine assignments

14) Minimizing flow time
15) Minimizing late jobs (number)/ lateness
16) Minimizing machine processing time
17) Minimizing production time
18) Filling the tool magazines as densely as possible
19) Maximizing assigned workload
20) Maximizing routing flexibility of batches
21) Maximizing the sum of operations priorities
22) Minimizing material handling time
23) Minimizing total distance travelled by parts during production
24) Minimizing total number of cutting tools required
25) Minimizing workload of machines
26) Minimizing breakdowns
27) Minimizing earliness

After regressive analysis of the loading objectives of various researchers the optimization approaches and

Table 3: Review of machine loading problems in FMS based on optimization techniques not falling in the above classification

OTHER OPTIMIZATION TECHNIQUES						
1	1984	K. E. Stecke & T. L. Morin [67]	Single server closed queueing network model	Ø Balancing of workload	Maximizes expected production of FMS	Results are compared and contrasted with previous models of production systems
2	1986	K. E. Stecke [68]	Hierarchical approach	Ø Maximizing throughput	Nonlinear integer programs models	Ties with some previous results & use of the proposed models to solve realistic loading problems is discussed
3	1986	J. G. Shanthikumar & K. E. Stecke [69]	Dynamic approach	Ø Balancing of workload	Result maximizes expected production	results obtained here complement previous results from literature
4	1993	Y.-D. Kim [70]	Due-Date Based Loading methods	Ø Maximizing throughput	Results reported reduced tardiness and makespan & increased throughput	Computational tests
5	1997	H.-K. Roh & Y-D. Kim [71]	Due-Date Based Loading methods	Ø Minimizing total tardiness	Iterative approach performs better than others	Computational tests on randomly generated problems
6	1997	D. H. Lee, S. K. Lim, G. C. Lee, H. B. Jun, & Y. D. Kim [72]	Iterative algorithms	Ø Minimizing subcontracting costs	Solved part selection and loading problems	computational experiments on randomly generated test problems
7	1997	Y. D. Kim and C. A. Yano [73]	Queueing network model	Ø Maximizing throughput Ø Maximizing make span Ø Balancing of workload	Reducing number of machine groups and balancing workloads among machines help to reduce make span	Computational results are produced

8	1998	D.-H. Lee & Y.-D. Kim [74]	Iterative procedures	Ø Minimizing earliness Ø Minimizing tardiness Ø Minimizing subcontracting costs	Computational experiments on randomly generated test problems are produced	computational experiments are done on randomly generated test problems and the results are compared with existing results
9	1999	J. N. D. Gupta, L. H. S. Luong, & V. H. Nguyen [75]	Dispatching approach	Ø Minimizing make spans Ø Minimizing average flow time Ø Minimizing tardiness	Satisfactory performance of given dispatching algorithm	Simulation results are compared with existing results
10	2000	S. Rahimifard & S.T. Newman [76]	Combined machine loading (CML) algorithms	Ø Meeting delivery dates Ø Minimising production costs	Adoption of algorithms within an application is dependent on number of manufacturing constraints	Computational results are produced and performance measure is carried out in virtual environment
11	2012	W. F. Mahmudy, R. M. Marian, & L. H. S. Luong [77]	Real coded genetic algorithms (RCGA)	Ø Maximizing throughput Ø Minimizing system unbalance	RCGA improves FMS performance & minimizes required computational time	Results are compared to the previous literature work

The tick marks (√) shows the density of repetitive occurrence of the optimization techniques and approaches for solving the machine loading problem.

Abbreviations used in Table 5:

1) Genetic Algorithm (GA): GA, Hybrid GA, Constraint based GA, Constraint-chromosome GA, Real coded GA, integrated approach based on GA
2) Heuristic Algorithm (HA): HA, Fast HA, Fuzzy based HA, GA based HA, Hybrid TS and SA based HA, Lagrangian based HA, GA and PSO based Meta-hybrid HA, multi stage programming approach based HA
3) Simulated annealing (SA): SA, Constraints-Based Fast SA, GA based SA, Hybrid GA-SA & SA-TS algorithm
4) Mathematical programming (MP): MP, Linear MP, Non-linear MP, GA based MP
5) Swarm Optimization (SO): SO, Particle SO (PSO), Modified PSO
6) Queueing network model (QNM): QNM, Single server closed QNM
7) Mixed-integer programming (MIP): MIP, GA based MIP
8) Branch and bound algorithms (B&BA) : B&BA, New B&BA
9) Integer programming (IP): IP, linear IP
10) Non-linear programming
11) Stochastic model

Table 4: Objectives of machine loading in FMS

Sr.	Year	Researcher Name	1	2	3	4	5	6	7	8	9	10	11	12	13	14	15	16	17	18	19	20	21	22	23	24	25	26	27
1	1983	K. E. Stecke & F. B. Talbot [2]			√						√	√																	
2	1984	K. E. Stecke & T. L. Morin [67]			√																								
3	1985	K. Shankar & Y. J. Tzen [14]	√		√												√												
4	1986	M. Berrada & K. E. Stecke [26]			√																								
5	1986	K. E. Stecke [68]		√																									
6	1986	J.G.S. Kumar & K. E. Stecke [69]			√																								
7	1986	T. J. Greene & R. Sadowski [29]				√										√	√												
8	1987	S.C. Sarin & C.S. Chen [30]								√																			
9	1988	J. A. Ventura et al. [15]				√																							
10	1989	K. Shankar & A. Srinivasulu [27]		√							√										√								
11	1990	B. Ram et al. [16]						√																					
12	1990	K. M. Bretthauer et al. [31]													√														
13	1990	H. C. Co et al. [32]			√																								
14	1990	M. Liang & S. P. Dutt [33]								√																			
15	1992	Y. D. Kim & C. A. Yano [28]		√																									
16	1992	S. K. Mukhopadhyay et al. [17]	√	√																									
17	1993	K. Kato et al. [18]	√																							√			
18	1993	Ming Liang [34]		√																									
19	1993	Y-D. Kim [70]		√																									
20	1994	Ming Liang [35]		√						√																			
21	1995	E. K. Steeke & F. B. Talbot [19]			√						√	√																	
22	1997	M. K. Tiwari et al. [20]	√	√																									
23	1997	V. N. Hsu & R. D. Matta [36]						√																					
24	1997	H.-K. Roh & Y.-D. Kim [71]							√																				
25	1997	D. H. Lee et al. [72]												√															
26	1997	Y. D. Kim and C. A. Yano [73]		√	√	√																							
27	1998	S. K. Mukhopadhyay et al. [44]	√																										
28	1998	D.-H. Lee & Y.-D. Kim [74]							√					√															√
29	1998	G. K. Nayak & D. Acharya [21]											√									√							
30	1998	T. J. Sawik [37]			√							√																	
31	1999	F. Guerrero et al. [38]			√																								
32	1999	J. N. D. Gupta et al. [75]				√			√							√													
33	2000	N. Kumar & K. Shanker [49]	√										√																
34	2000	D.-H. Lee & Y.-D. Kim [22]																									√		
35	2000	S. Rahimifard & S. Newman [76]					√			√																			

Table 5: Optimization techniques used for solving machine loading problems in FMS

36	2000	M. K. Tiwari & N. Vidyarthi [55]	√	√																								
37	2001	N. Kumar & K. Shanker [39]			√																							
38	2001	N. K. Vidyarthi & M. K. Tiwari [64]	√	√																								
39	2002	H. Yong & Z. Wu [50]			√																							
40	2003	M. Gamila & S. Motavalli [40]															√						√					
41	2004	R. R. Kumar et al. [65]	√	√																								
42	2004	T. Sawik [41]																√										
43	2004	R. Swarnkar & M. K. Tiwari [45]	√	√																								
44	2005	M. Aldaihani & M. Savsar [46]						√																				
45	2006	N. Nagarjunaa et al. [23]	√																									
46	2006	M. Goswami & M. Tiwari [24]	√	√																								
47	2006	M. K. Tiwari et al. [47]	√	√																								
48	2006	A. Kumar et al. [51]			√						√	√						√		√			√					
49	2007	A. Turkcan et al. [52]						√	√																			
50	2007	M. K. Tiwari et al. [25]	√	√																								
51	2007	S. Biswas & S. Mahapatra [61]	√																									
52	2008	A. Prakash et al. [66]	√	√																								
53	2008	S. Biswas & S. Mahapatra [62]	√																									
54	2008	S. Ponnambalam & L. Kiat [63]	√	√																								
55	2008	V. Tyagi & A. Jain [53]	√																									
56	2009	M. Yogeswaran et al. [56]	√	√																								
57	2010	S. K. Mandal et al. [57]	√	√		√																					√
58	2011	M. I. Mgwatua [42]		√		√																						
59	2012	V. M Kumar et al. [58]	√	√																								
60	2012	C. Basnet [59]	√																									
61	2012	M. Arıkan & S. Erol [48]	√		√										√													
62	2012	U. K. Yusof et al. [54]	√	√																								
63	2012	D. Kosucuoglu & U. Bilge [60]																						√				
64	2012	A. M. Abazari et al. [43]	√																									
65	2012	W. F. Mahmudy et al. [77]	√	√																								

12) Modified immune algorithm
13) Approximative lexicographic approach
14) Iterative algorithms
15) Hierarchical approach
16) Branch & backtrack procedure
17) Combined machine loading algorithms
18) Dispatching approach
19) Due-Date Based Loading methods

20) Dynamic approach
21) Fuzzy Logic

CONCLUSION ARRIVED ON MACHINE LOADING OBJECTIVES AND OPTIMIZATION TECHNIQUES IN FMS

Detailed study of the machine loading problem is conducted by the authors. The conclusions of the research throttled are divided into three sections as below.

Conclusion on Machine Loading Objectives

On exhaustive study, twenty eight loading objectives are observed in the reviewed literature. Tick marks (√) in Table 4 are showing the density for repeatability of the machine loading objectives, which concludes that a research with maximum loading objectives is still required for solving the machine problem. Maximizing expected production rate (throughput) & unbalancing the workload per machine for a system of groups of pooled machines of unequal sizes are the two objectives on which most of the researchers have worked. Balancing of workload on machines for a system of groups of pooled machines of equal sizes is the second most researched loading objective. Minimizing make span is the third most researched loading objective. Minimizing job tardiness is fourth loading objective in the order. Minimizing mean job flow time & minimizing production cost are found at fifth position in the order. Loading objectives observed at sixth rank are maximizing profitability, maximizing the assigned workload, maximizing the part types in each batch, maximizing utilization of system, minimizing subcontracting costs and minimizing the total number of cutting tools required. Material handling time, maximizing routing flexibility of the batches, minimizing earliness, minimizing mean lateness, minimizing mean machine idle time, minimizing overall machining cost, minimizing production time, minimizing the effect of breakdowns, minimizing the maximum workload of the machines, minimizing the number of late jobs, minimizing total flexible manufacturing cell cost per unit of production, minimizing total inter-station transfer time, minimizing total processing time, minimizing part movements and minimisation of the total distance travelled by parts during their production are the loading objectives that are least considered.

Conclusion on Optimization Techniques in FMS

The categorized literature review concludes that the researcher's major emphasis and contribution are towards the use and application of global optimization techniques and with natural optimization techniques, too. Heuristic Algorithms is the mostly used optimization technique by researchers, followed by Genetic Algorithms (GA). Mixed Integer Programming (MIP) & Simulated Annealing (SA) approach are the third mostly used optimization techniques. Linear Mathematical Programming (LMP) is next in the queue succeeded by Integer Programming (IP). At sixth level is Particle Swarm Optimization (PSO) approach. The least used optimization techniques are Tabu search, Swarm Optimization Approach, Branch and backtrack procedure, Branch and bound approach, Combined machine loading (CML) algorithms, Dispatching approach, Due-Date Based methods, Dynamic approach, Fuzzy Logic, Global criterion approach, Hierarchical approach, Artificial immune algorithm, Iterative algorithms, Lexicographic approach, Non-linear programming, Queueing network model and Stochastic model.

Conclusion on validation approaches

A few research problems are solved and the results are compared with previous research results. The results are validated by comparing with literature available results.

Methodologies Findings and Interpretations

A problem when solved for a limited or less number of objectives, it is rather a customized solution for a problem. For general solution, the problem needs to be solved for all possible objectives. On extreme analysis of the machine loading problem and objectives, and on discussion with the academicians and industrialists, the authors emphasise to solve the loading problem for maximization of throughput, part types in a batch, routing flexibility, balancing/unbalancing of system and workload, and minimization of make-span, delivery dates (covering lateness, tardiness and earliness), part movements, subcontracting costs, machine processing time, tool magazine capacity, number of cutting tools required, breakdowns, non-splitting of jobs, time spend by job on machines in one study. Machine loading problem should be solved for general solution to the problem, for maximum number of objectives. All these objectives are having a common goal of optimizing the production and manufacturing costs.

The literature review reports the application of heuristics, global optimization techniques and some other optimization techniques for solving the loading problem for the listed objectives. Among these approaches, the global optimization techniques were more frequently adopted and the results as founded by the researchers were more accurate and acceptable. Based on regressive analysis of the available literature, and skills and concluding remarks, the authors suggest for the use of natural optimization techniques like swarm optimization for further research. The results of swarm optimization were found more reliable and acceptable as compared to GA, and PSO has attractive characteristics. PSO retains knowledge of all previous particles, which is destroyed in GA when the population changes. PSO is a mechanism of constructive cooperation and information-sharing between particles. Due to the simple concept, ease of implementation, and quick convergence, PSO has gained much attention and has been successfully applied to a wide range of applications.

RESEARCH GAPS AND SCOPE OF RESEARCH IN LOADING OF MACHINES IN FMS

There exists a research gap among the literature available. There are several future scopes that are still not worked out, or still to be worked in a more optimized manner. Based on our observation and exhaustive study such revealed research gap are listed below: Need of integration of loading with other decisions in the neighbourhood of loading (K. Shankar & A. K. Agrawal, 1991); need to reduce excessive computing times (Y. D. Kim & C. A. Yano, 1989); further need of optimization (N. K. Vidyarthi & M. K. Tiwari, 2001, M. K. Tiwari et al., 2007; Amir Musa Abazari et al., 2012); research is required to develop planning softwares (D. H. Lee et al., 1997); PLC controller needs to be enhanced (M. C. Zhou et al., 1993); waiting time for parts and idling time for machines need attention [Mussa I. Mgwatu, 2011]; research by imposing constraints on the availability of resources i.e. jigs, fixtures, pallets, material handling devices needs to be carried out (K. Kato, 1993, N. K. Vidyarthi & M. K. Tiwari, 2001; N. Nagarjuna et al., 2006; Akhilesh Kumar et al., 2006; M. K. Tiwari et al., 2007; Sandhyarani Biswas & S. S. Mahapatra, 2007; Sandhyarani Biswas & S. S. Mahapatra, 2008; Santosh Kumar Mandal et al. 2010; Amir Musa Abazari et al., 2012); new solution methodology needs to be proposed (Santosh Kumar Mandal et al. 2010); need of AI in the field of FMS) Chinyao Low et al., 2006; Sandhyarani Biswas & S. S. Mahapatra, 2008); Need to use dedicated robot (Majid M. Aldaihani & Mehmet Savsar, 2005); need of simulation studies for FMS (K. Shankar & A. K. Agrawal, 1991; N. K. Vidyarthi & M. K. Tiwari, 2001). Availability of a number of research gaps and that too identified by various eminent researchers from time to time evacuates the need of vast research for solving the observed PPC problems i.e. machine loading problems in FMS.

The authors are working to solve the loading problem with more number of objectives in a single study and for the development of knowledge base system for the machine loading problem. The authors suggest for the development of a knowledge base for all five productions planning problems; part type selection problem, machine grouping problem, production ratio problem, resource allocation problem and loading problem in a single study incorporating the individual objectives of the five individual problems and their respective technological and capacity constraints.

REFERENCES

1. Stecke, K.E. (1983) Formulation and Solution of Nonlinear Integer Production Planning Problems for Flexible Manufacturing Systems. Management Science, 29, 273-288. http://dx.doi.org/10.1287/mnsc.29.3.273
2. Stecke, K.E. and Talbot, F.B. (1983) Heuristic Loading Algorithms for Flexible Manufacturing Systems. Proceedings of the Seventh International Conference on Production Research, Windsor, 22-24 August 1983.
3. Muhlenbein, H. (1992) Parallel Genetic Algorithms in Combinatorial Optimization. In: Balci, O., Sharda, R. and Zenios, S.A., Eds., Computer Science and Operations Research: New Developments in Their Interfaces, Pergamon Press, Oxford, 441-456.
4. Russell, S.J. and Norvig, P. (2002) Artificial Intelligence: A Modern Approach. Second Edition, Prentice Hall, Englewood Cliffs.
5. Rastrigin, L.A. (1963) The Convergence of the Random Search Method in the External Control of Many-Parameter System. Automation and Remote Control, 24, 1337-1342.
6. Metropolis, N., Rosenbluth, A.W., Rosenbluth, M.N., Teller, A.H. and Teller, E. (1953) Equation of State Calculations by Fast Computing Machines. The Journal of Chemical Physics, 21, 1087-1092. http://dx.doi.org/10.1063/1.1699114
7. Glover, F. (1989) Tabu Search—Part I. Operations Research Society of America (ORSA). Journal on Computing, 1, 90-206. http://dx.doi.org/10.1287/ijoc.1.3.190
8. Moscato, P. (1989) On Evolution, Search, Optimization, Genetic Algorithms and Martial Arts: Towards Memetic Algorithms. Technical Report C3P 826, Caltech Con-Current Computation Program 158-79, California Institute of Technology, Pasadena.
9. Holland, J.H. (1962) Outline for a Logical Theory of Adaptive Systems. Journal of the ACM, 9, 297-314. http://dx.doi.org/10.1145/321127.321128
10. Holland, J.H. and Reitman, J.S. (1977) Cognitive Systems Based on Adaptive Algorithms. ACM SIGART Bulletin, 63, 49.

11. http://dx.doi.org/10.1145/1045343.1045373
12. Friedberg, R.M. (1958) A Learning Machine: Part I. IBM Journal of Research and Development, 2, 2-13.
13. http://dx.doi.org/10.1147/rd.21.0002
14. Dorigo, M., Maniezzo, V. and Colorni, A. (1996) The Ant System: Optimization by a Colony of Cooperating Agents. IEEE Transactions on Systems, Man, and Cybernetics Part B: Cybernetics, 26, 29-41.
15. http://dx.doi.org/10.1109/3477.484436
16. Eberhart, R.C. and Kennedy, J. (1995) A New Optimizer Using Particle Swarm Theory. Proceedings of the Sixth International Symposium on Micro Machine and Human Science, Nagoya, 4-6 October 1995, 39-43.
17. http://dx.doi.org/10.1109/MHS.1995.494215
18. Shankar, K. and Tzen, Y.J.J. (1985) A Loading and Dispatching Problem in a Random Flexible Manufacturing System. International Journal of Production Research, 23, 579-595.
19. http://dx.doi.org/10.1080/00207548508904730
20. Ventura, J.A., Chen, F.F. and Leonard, M.S. (1988) Loading Tools to Machines in Flexible Manufacturing Systems. Computers & Industrial Engineering, 15, 223-230.
21. Ram, B., Sarin, S. and Chen, C.S. (1990) A Model and Solution Approach for the Machine Loading and Tool Allocation Problem in FMS. International Journal of Production Research, 28, 637-645.
22. Mukhopadhyay, S.K., Midha, S. and Murlikrishna, V. (1992) A Heuristic Procedure for Loading Problem in Flexible Manufacturing Systems. International Journal of Production Research, 30, 2213-2228.
23. http://dx.doi.org/10.1080/00207549208948146
24. Kato, K., Oba, F. and Hashimoto, F. (1993) Loading and Batch Formation in Flexible Manufacturing Systems. Control Engineering Practice, 1, 845-850.
25. http://dx.doi.org/10.1016/0967-0661(93)90252-M
26. Steeke, E.K. and Talbot, F.B. (1995) Heuristics for Loading Flexible Manufacturing Systems, Flexible Manufacturing Systems: Recent Developments. Elsevier Science B.V., Amsterdam, 171-176.
27. Tiwari, M.K., Hazarika, B., Vidyarthi, N.K., Jaggi, P. and Mukhopadhyay, S.K. (1997) A Heuristic Solution Approach to the Machine Loading Problem of FMS and Its Petri Net Model. International Journal of Production Research, 35, 2269-2284.

28. http://dx.doi.org/10.1080/002075497194840
29. Nayak, G.K. and Acharya, A.D. (1998) Part Type Selection, Machine Loading and Part Type Volume Determination in FMS Planning. International Journal of Production Research, 36, 1801-1824.
30. http://dx.doi.org/10.1080/002075498192977
31. Lee, D.H. and Kim, Y.-D. (2000) Loading Algorithms for Flexible Manufacturing Systems with Partially Grouped Machines. IIE Transactions, 32, 33-47.
32. Nagarjuna, N., Mahesh, O. and Rajagopal, K. (2006) A Heuristic Based on Multi-Stage Programming Approach for Machine-Loading Problem in a Flexible Manufacturing System. Robotics and Computer-Integrated Manufacturing, 22, 342-352.
33. http://dx.doi.org/10.1016/j.rcim.2005.07.006
34. Goswami, M. and Tiwari, M.K. (2006) A Reallocation-Based Heuristic to Solve a Machine Loading Problem with Material Handling Constraint in a Flexible Manufacturing System. International Journal of Production Research, 44, 569-588.
35. Tiwari, M.K., Saha, J. and Mukhopadhyay, S.K. (2007) Heuristic Solution Approaches for Combined-Job Sequencing and Machine Loading Problem in Flexible Manufacturing Systems. International Journal of Advanced Manufacturing Technology, 31, 716-730.
36. Berrada, M. and Stecke, K.E. (1986) A Branch and Bound Approach for Machine Load Balancing in Flexible Manufacturing Systems. Management Science, 32, 1316-1335.
37. http://dx.doi.org/10.1287/mnsc.32.10.1316
38. Shankar, K. and Srinivasulu, A. (1989) Some Selection Methodologies for Loading Problems in a Flexible Manufacturing System. International Journal of Production Research, 27, 1019-1034.
39. http://dx.doi.org/10.1080/00207548908942605
40. Kim, Y.-D. and Yano, C.A. (1994) A New Branch and Bound Algorithm for Loading Problems in Flexible Manufacturing Systems. International Journal of Flexible Manufacturing Systems, 6, 361-381.
41. http://dx.doi.org/10.1007/BF01324801
42. Greene, T.J. and Sadowski, R.P. (1986) A Mixed Integer Programming for Loading and Scheduling Multiple Manufacturing Cells. European Journal of Operation Research, 24, 379-386.
43. http://dx.doi.org/10.1016/0377-2217(86)90031-7

44. Sarin, S.C. and Chen, C.S. (1987) The Machine Loading and Tool Allocation Problem in a Flexible Manufacturing System. International Journal of Production Research, 25, 1081-1094.
45. http://dx.doi.org/10.1080/00207548708919897
46. Bretthauer, K.M. and Venkataramanan, M.A. (1990) Machine Loading and Alternate Routing in a Flexible Manufacturing System. Computers and Industrial Engineering, 18, 341-350.
47. http://dx.doi.org/10.1016/0360-8352(90)90056-R
48. Co, H.C., Biermann, J.S. and Chen, S.K. (1990) A Methodical Approach to the Flexible Manufacturing System Batching, Loading and Tool Configuration Problems. International Journal of Production Research, 28, 2171-2186.
49. http://dx.doi.org/10.1080/00207549008942860
50. Liang, M. and Dutt, S.P. (1990) A Mixed-Integer Programming Approach to the Machine Loading and Process Planning Problem in a Process Layout Environment. International Journal of Production Research, 28, 1471-1484.
51. http://dx.doi.org/10.1080/00207549008942806
52. Liang, M. (1993) Part Selection, Machine Loading and Machining Speed Selection in Flexible Manufacturing Systems. Computers and Industrial Engineering, 25, 259-262.
53. http://dx.doi.org/10.1016/0360-8352(93)90270-8
54. Liang, M. (1994) Integrating Machining Speed, Part Selection and Machine Loading Decisions in Flexible Manufacturing Systems. Computers & Industrial Engineering, 26, 599-608.
55. Hsu, V.N. and De Matta, R. (1997) An Efficient Heuristic Approach to Recognize the Infeasibility of a Loading Problem. International Journal of Manufacturing Systems, 9, 31-50.
56. Sawik, T.J. (1998) A Lexicographic Approach to Bi-Objective Loading of a Flexible Assembly System. European Journal of Operational Research, 107, 656-668. http://dx.doi.org/10.1016/S0377-2217(97)00091-X
57. Guerreore, F., Lozano, S., Koltai, T. and Larraneta, J. (1999) Machine Loading and Part Type Selection in Flexible Manufacturing System. International Journal of Production Research, 37, 1303-1317. http://dx.doi.org/10.1080/002075499191265
58. Kumar, N. and Shanker, K. (2001) Comparing the Effectiveness of Workload Balancing Objectives in FMS Loading. International Journal of Production Research, 39, 843-871.

59. Gamila, M.A. and Motavalli, S. (2003) A Modeling Technique for Loading and Scheduling Problems in FMS. Robotics and Computer Integrated Manufacturing, 19, 45-54.

60. Sawik, T. (2004) Loading and Scheduling of a Flexible Assembly System by Mixed Integer Programming. European Journal of Operational Research, 154, 1-19. http://dx.doi.org/10.1016/S0377-2217(02)00795-6

61. Mgwatua, M.I. (2011) Interactive Decisions of Part Selection, Machine Loading, Machining Optimisation and Part Scheduling Sub-Problems for Flexible Manufacturing Systems. International Transaction Journal of Engineering, Management, & Applied Sciences & Technologies, 2, 93-109.

62. Abazari, A.M., Solimanpur, M. and Sattari, H. (2012) Optimum Loading of Machines in a Flexible Manufacturing System Using a Mixed-Integer Linear Mathematical Programming Model and Genetic Algorithm. Computers & Industrial Engineering, 62, 469-478. http://dx.doi.org/10.1016/j.cie.2011.10.013

63. Mukhopadhyay, S.K., Singh, M.K. and Srivastava, R. (1998) FMS Loading: A Simulated Annealing Approach. International Journal of Production Research, 36, 1529-1547. http://dx.doi.org/10.1080/002075498193156

64. Swarnkar, R. and Tiwari, M.K. (2004) Modeling Machine Loading Problem of FMSs and Its Solution Methodology Using a Hybrid Tabu Search and Simulated Annealing-Based Heuristic Approach. Robotics and Computer-Integrated Manufacturing, 20, 199-209. http://dx.doi.org/10.1016/j.rcim.2003.09.001

65. Aldaihani, M.M. and Savsar, M. (2005) A Stochastic Model for the Analysis of a Two-Machine Flexible Manufacturing Cell. Computers & Industrial Engineering, 49, 600-610. http://dx.doi.org/10.1016/j.cie.2005.09.002

66. Tiwari, M.K., Kumar, S., Kumar, S., Prakash and Shankar, R. (2006) Solving Part-Type Selection and Operation Allocation Problems in an FMS: An Approach Using Constraints-Based Fast Simulated Annealing Algorithm. IEEE Transactions on Systems, Man, and Cybernetics—Part A: Systems and Humans, 36, 1170-1184.

67. Arikan, M. and Erol, S. (2012) A Hybrid Simulated Annealing-Tabu Search Algorithm for the Part Selection and Machine Loading Problems in Flexible Manufacturing Systems. International Journal of Advanced Manufacturing Technology, 59, 669-679. http://dx.doi.org/10.1007/s00170-011-3506-0

68. Kumar, N. and Shanker, K. (2000) A Genetic Algorithm for FMS Part Type Selection and Machine Loading. International Journal of Production Research, 38, 3861-3887.
69. Yong, H.H. and Wu, Z.M. (2002) GA-Based Integrated Approach to FMS Part Type Selection and Machine Loading Problem. International Journal of Production Research, 40, 4093-4110. http://dx.doi.org/10.1080/00207540210146972
70. Kumar, A., Prakash, Tiwari, M.K., Shankar, R. and Baveja, A. (2006) Solving Machine-Loading Problem of a Flexible Manufacturing System with Constraint-Based Genetic Algorithm. European Journal of Operational Research, 175, 1043-1069. http://dx.doi.org/10.1016/j.ejor.2005.06.025
71. Turkcan, A., Akturk, M.S. and Storer, R.H. (2007) Due Date and Costbased FMS Loading, Scheduling and Tool Management. International Journal of Production Research, 45, 1183-1213.
72. Tyagi, V. and Jain, A. (2008) Assessing the Effectiveness of Flexible Process Plans for Loading and Part Type Selection in FMS. Advances in Production Engineering & Management, 3, 27-44.
73. Yusof, U.K., Budiarto, R. and Deris, S. (2012) Constraint-Chromosome Genetic Algorithm for Flexible Manufacturing System Machine-Loading Problem. International Journal of Innovative Computing, Information and Control, 8, 1591-1609.
74. Tiwari, M.K. and Vidyarthi, N.K. (2000) Solving Machine Loading Problem in Flexible Manufacturing System Using Genetic Algorithm Based Heuristic Approach. International Journal of Production Research, 38, 3357-3384. http://dx.doi.org/10.1080/002075400418298
75. Yogeswaran, M., Ponnambalam, S.G. and Tiwari, M.K. (2009) An Efficient Hybrid Evolutionary Heuristic Using Genetic Algorithm and Simulated Annealing Algorithm to Solve Machine Loading Problem in FMS. International Journal of Production Research, 47, 5421-5448.
76. Mandal, S.K., Pandey, M.K. and Tiwari, M.K. (2010) Incorporating Dynamism in Traditional Machine Loading Problem: An AI-Based Optimization Approach. International Journal of Production Research, 48, 3535-3559. http://dx.doi.org/10.1080/00207540902814306
77. Kumar, V.M., Murthy, A.N.N. and Chandrashekar, K. (2012) A Hybrid Algorithm Optimization Approach for Machine Loading Problem in Flexible Manufacturing System. Journal of Industrial Engineering International, 8, 3. http://dx.doi.org/10.1186/2251-712X-8-3

78. Basnet, C. (2012) A Hybrid Genetic Algorithm for a Loading Problem in Flexible Manufacturing Systems. International Journal of Production Research, 50, 707-718.
79. Kosucuoglu, D. and Bilge, U. (2012) Material Handling Considerations in the FMS Loading Problem with Full Routing Flexibility. International Journal of Production Research, 50, 6530-6552.
80. Biswas, S. and Mahapatra, S.S. (2007) Machine Loading in Flexible Manufacturing System: A Swarm Optimization Approach. Proceedings of the Eighth International Conference on Operations and Quantitative Management, Bangkok, 17-20 October 2007.
81. Biswas, S. and Mahapatra, S.S. (2008) Modified Particle Swarm Optimization for Solving Machine Loading Problems in Flexible Manufacturing Systems. International Journal of Advanced Manufacturing Technology, 39, 931-942.
82. Ponnambalam, S.G. and Kiat, L.S. (2008) Solving Machine Loading Problem in Flexible Manufacturing Systems Using Particle Swarm Optimization. World Academy of Science, Engineering and Technology, 39, 14-19.
83. Vidyarthi, N.K. and Tiwari, M.K. (2001) Machine Loading Problem of FMS: A Fuzzy-Based Heuristic Approach. International Journal of Production Research, 39, 953-979. http://dx.doi.org/10.1080/00207540010010244
84. Kumar, R.R., Singh, A.K. and Tiwari, M.K. (2004) A Fuzzy Based Algorithm to Solve the Machine-Loading Problems of a FMS and Its Neuro Fuzzy Petri Net Model. International Journal of Advanced Manufacturing Technology, 23, 318-341.
85. http://dx.doi.org/10.1007/s00170-002-1499-4
86. Prakash, A., Khilwani, N., Tiwari, M.K. and Cohen, Y. (2008) Modified Immune Algorithm for Job Selection and Operation Allocation Problem in Flexible Manufacturing Systems. Advances in Engineering Software, 39, 219-232. http://dx.doi.org/10.1016/j.advengsoft.2007.01.024
87. Stecke, K.E. and Morin, T.L. (1985) The Optimality of Balancing Workloads in Certain Types of Flexible Manufacturing Systems. European Journal of Operational Research, 20, 68-82.
88. Stecke, K.E. (1986) A Hierarchical Approach to Solving Grouping and Loading Problems of Flexible Manufacturing Systems. European Journal of Operational Research, 24, 369-378. http://dx.doi.org/10.1016/0377-2217(86)90030-5
89. Shanthikumar, J.G. and Stecke, K.E. (1986) Reducing Work in Progress Inventory in Certain Classes of Flexible Manufacturing Systems.

European Journal of Operation Research, 26, 266-271. http://dx.doi.org/10.1016/0377-2217(86)90189-X

90. Kim, Y.-D. (1993) A Study on Surrogate Objectives for Loading a Certain Type of Flexible Manufacturing Systems. International Journal of Production Research, 31, 381-392. http://dx.doi.org/10.1016/0377-2217(86)90189-X
91. Roh, H.-K. and Kim, Y.-D. (1997) Due-Date Based Loading and Scheduling Methods for a Flexible Manufacturing System with an Automatic Tool Transporter. International Journal of Production Research, 35, 2989-3004.
92. Lee, D.-H., Lira, S.-K., Lee, G.-C., Jun, H.-B. and Kim, Y.-D. (1997) Multi-Period Part Selection and Loading Problems in Flexible Manufacturing Systems. Computers & Industrial Engineering, 33, 541-544.
93. Kim, Y.D. and Yano, C.A. (1997) Impact of Throughput Based Objective and Machine Grouping Decisions on the Short-Term Performance of Flexible Manufacturing System. International Journal of Production Research, 35, 3303-3322. http://dx.doi.org/10.1080/002075497194084
94. Lee, D.-H. and Kim, Y.-D. (1998) Iterative Procedures for Multi-Period Order Selection and Loading Problems in Flexible Manufacturing Systems. International Journal of Production Research, 36, 2653-2668. http://dx.doi.org/10.1080/002075498192418
95. Gupta, J.N.D. (1999) Part Dispatching and Machine Loading in Flexible Manufacturing System Using Central Queues. International Journal of Production Research, 37, 1427-1435. http://dx.doi.org/10.1080/002075499191337
96. Rahimifard, S. and Newman, S.T. (2000) Machine Loading Algorithms for the Elimination of Tardy Jobs in Flexible Batch Machining Applications. Journal of Materials Processing Technology, 107, 450-458.
97. Mahmudy, W.F., Marian, R.M. and Luong, L.H.S. (2012) Solving Part Type Selection and Loading Problem in Flexible Manufacturing System Using Real Coded Genetic Algorithms—Part II: Optimization. World Academy of Science, Engineering and Technology, 69, 778-782.

Chapter 6

A PREDICTION METHOD FOR IN-PLANE PERMEABILITY AND MANUFACTURING APPLICATIONS IN THE VARTM PROCESS

Ya-Jung Lee[1], Yu-Ti Jhan[1], Cheng-Hsien Chung[2], Yao Hsu[3]

[1]Engineering Science and Ocean Engineering, National Taiwan University, Chinese Taipei

[2]R & D Department, United Ship Design & Development Center, Chinese Taipei

[3]Business and Entrepreneurial Management, Kai Nan University, Taoyuan, Chinese Taipei

ABSTRACT

Vacuum Assisted Resin Transfer Molding (VARTM) is a popular method for manufacturing large-scaled, single-sided mold composite structures, such as wind turbine blades and yachts. Simulation to find the proper infusion scenario before manufacturing is essential to avoid dry spots as well as incomplete saturation and various fiber weaves with different permeability affect numerical simulation tremendously. This study focused on deriving the in-plane permeability prediction method for Fiber Reinforced Plastics (FRP) laminates in the VARTM process by experimental measurements and numerical analysis. The method provided an efficient way to determine the permeability of laminates without conducting lots of experiments in the future. In-plane permeability imported into the software, RTM-Worx, to simulate resin flowing pattern before the infusion experiments of a 3D ship hull with two different infusion scenarios. The close agreement between experiments and simulations proved the correctness and applicability of the prediction method for the in-plane permeability.

INTRODUCTION

Vacuum Assisted Rein Transfer Molding (VARTM) has been employed successfully to manufacture FRP (Fiber Reinforced Plastics) ship hull structures since 1994 [1] and other large structures, such as wind turbine blades and yachts [2-3]. In this manufacturing technique, the fiber lay-up is placed in advance above a one-sided mold, and resin injects into the fiber lay-up under one pressure gradient as vacuum is applied. This procedure provides the advantages of low VOC (Volatile Organic Compounds) emission as well as stable and excellent mechanical properties of products. The manufacturing failure risks, such as void contain and incomplete saturation, increase as the structure dimension increases. Accordingly, resin flowing prediction before manufacturing complex structures with various fiber laminates is essential. The CV-FEM (control volume finite element method) is utilized to explicate the behavior of resin flowing in fiber laminates [4,5]. With respect to the VARTM simulation, Mohan [6] modeled flow in channels based on a finite element method. Sun [7] and Ni [8] applied CV-FEM to predict the filling time and pattern associated with high-permeable medium and grooves in the SCRIMP process. Koorevaar [9,10] adopted 2.5D flow models, meaning that resin flow in the thickness direction is neglected, although the geometries are described as 3D, to simulate the VARTM and RTM process. The VARTM simulation involves with parameters, such as the viscosity of resin, the porosity and permeability of fiber laminates. The permeability of FRP laminates is critical in the manufacturing process and numerous researchers have investigated the measurements and analytical models of permeability [7,8] [11-15]. Sun [7] and Ni [8] set up a device to visualize the experimental results and measurements of permeability. Wang and Demaria [11-13] characterized the in-plane permeability of woven fabrics and proposed a predictive model. Wu [14] and Nedanov [15] researched the trans-plane permeability by theory and experiments to describe 3D permeability of fiber laminates. Some researchers have described the flow using equivalent permeability [16-18]. For example, Dong [18] increased the thickness of the distribution medium and/or the fiber preform to obtain equivalent material permeability. Besides, the Kozeny-Carman equation was commonly utilized to discuss the flow impregnates inside the porous material when considering the dimension of fiber [19-22]. Gebart [19] investigated the permeability of an idealized unidirectional reinforcement through the Darcy's law and Kozeny-Carman equation, and Rahatekar [21] used the permeability models through Kozeny-Carman equation to discuss the relationship between injection pultrusion pressure and process control parameters.

The infusion scenario in the VARTM process including the numbers of flowing channels, injection and venting ports, and the triggering timing of the injection gates etc. determines the impregnation of FRP products. Precise evaluation of the resin flowing pattern before fabrication reduces the risk and cost of manufacturing failure. Therefore, some researchers [23, 24] focused on how to arrange and improve the infusion process to avoid dry spots and incompletely saturation. Han [25] and Kang [26] analyzed the VARTM process under different vacuum conditions as well as infusion strategy and simulated the infusion process of a boat hull. The abovementioned studies have provided feasible methods for measuring permeability, but permeability measurement through the experiments is a time-consuming procedure. Hence, the development of simplified macroscopic models for predicting permeability of FRP laminates composing of various woven fabrics is beneficial for simulating and discussing the manufactures of complex structures with various infusion scenarios in the VARTM process. This study investigated the in-plane permeability for single and multi-types of fiber laminates by the permeability experiments. A prediction method for the in-plane permeability was established by defining the parameter of "thickness porosity", and the method provided an efficient way to determine the in-plane permeability of FRP laminates without conducting time-consuming experiments. Flowing patterns of a 3D ship hull with two different infusion scenarios were simulated to compare with the manufacturing experiments. The close agreement between experiments and simulations proved the correctness and applicability of the prediction method for the in-plane permeability.

IN-PLANE PERMEABILITY EXPERIMENTS

Experimental Measurements and Discussion

Several scholars have utilized Equation (1), derived from Darcy's Law, to describe the resin flowing process inside the porous space of fabrics under the RTM or VARTM method from 1980 to date.

$$V = -\frac{\overline{\overline{K}}}{\mu} \cdot \nabla P$$

$$\begin{bmatrix} u \\ v \\ w \end{bmatrix} = -\frac{1}{\mu} \begin{bmatrix} K_{xx} & K_{xy} & K_{xz} \\ K_{yx} & K_{yy} & K_{yz} \\ K_{zx} & K_{zy} & K_{zz} \end{bmatrix} \begin{bmatrix} \frac{\partial P}{\partial x} \\ \frac{\partial P}{\partial y} \\ \frac{\partial P}{\partial z} \end{bmatrix} \quad (1)$$

where V represents the Darcian Velocity, $\overline{\overline{K}}$ is the tensor of permeability, μ is the viscosity of resin, and ∇P is the gradient of pressure. In the research, realizing the in-plane permeant characteristic of FRP laminates was the main objective, and the length and width scale of fiber laminates was relatively larger than its thickness. Hence, the governing Equation (1) transformed into a 1D flowing Equation (2) when operating the rectangle experimental arrangement.

$$\frac{K}{\phi} = \frac{\mu}{2\Delta P}\frac{L^2}{t} \qquad (2)$$

where ΔP is the pressure gradient, ϕ is the porosity of laminate, L is the flow front, and t is the infusion time. Equation (2) reveals that the combination of value L^2/t, evaluating from the video records of flowing processes, the porosity of fiber laminates, and resin viscosity derives the in-plane permeability (K).

The viscosity of polyester resin, μ, changing with temperature (**Figure 1**(a)) was measured by the Brookfield viscometer and the thermostat sink (**Figure 1**(b)). The infusion time of the experimental process did not exceed 30 minutes and the ambient temperature did not change sharply during the infusion time. Accordingly, the viscosity was reasonably regarded as constant throughout the whole process. **Figure 2**(a) displays the in-plane permeability experimental schema. The fiber assembly was rectangular, and the dimensions of fiber laminates were 48 cm × 16 cm. Pressure gradient designed to be along the length of fiber laminates, from injection gate to venting port. The extensible spring linking with the injection gate was placed in the starting position to transform a point-injection into a line-injection, and consequently resin flowing front was perpendicular to the infusion direction (**Figure 2**(b)). **Table 1** presents the single-type fiber laminates under consideration, including strand chopped mat, woven roving and axial fabrics with different area density and numbers of layers. **Figure 3** shows the relationship between flowing fronts and its consuming infusion time which determines the permeability of FRP laminates. Initial time-delay region on the bottom surface was neglected due to the transverse-plane infusion from top surface to bottom surface. **Figure 4** shows the experimental permeability results of single-type laminates, and the horizontal axis represented the experimental laminates and the vertical axis was the in-plane permeability. The right-most bar symbolizing of D + P was the assembly consisted only with distribution medium and peel ply, and tested its permeability by considering the nestling effect [7], which was about the effect of fiber layers inserting mutually. Resin flowed more rapidly than other laminates because no fiber ply needed to be impregnated during the infusion process and the permeability, 1.88×10^{-9} m^2, was the highest between these results.

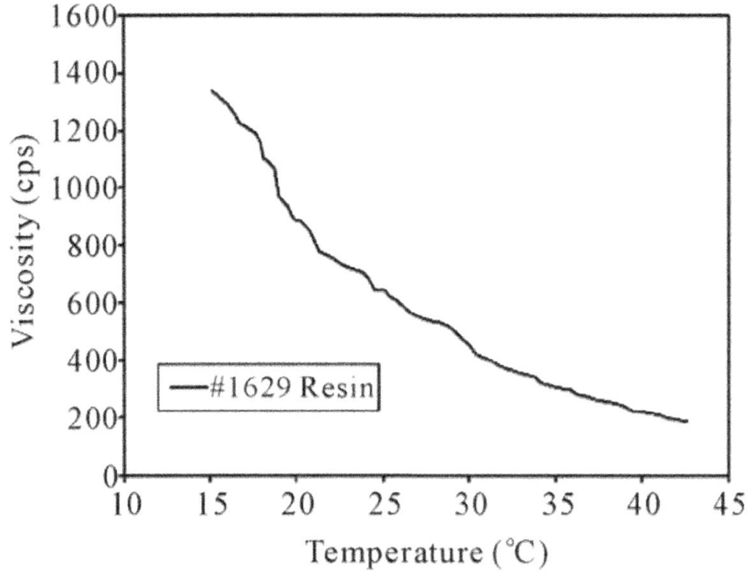

(a)

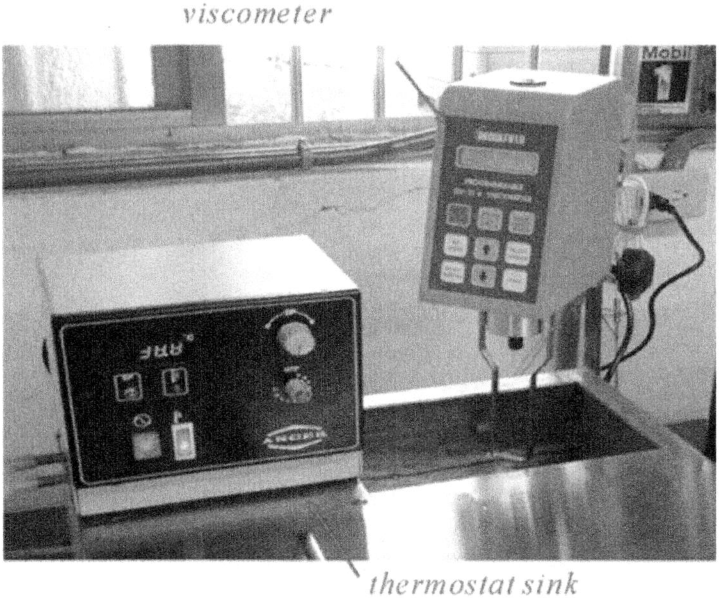

(b)

Figure 1: Viscosity measurement of polyester resin

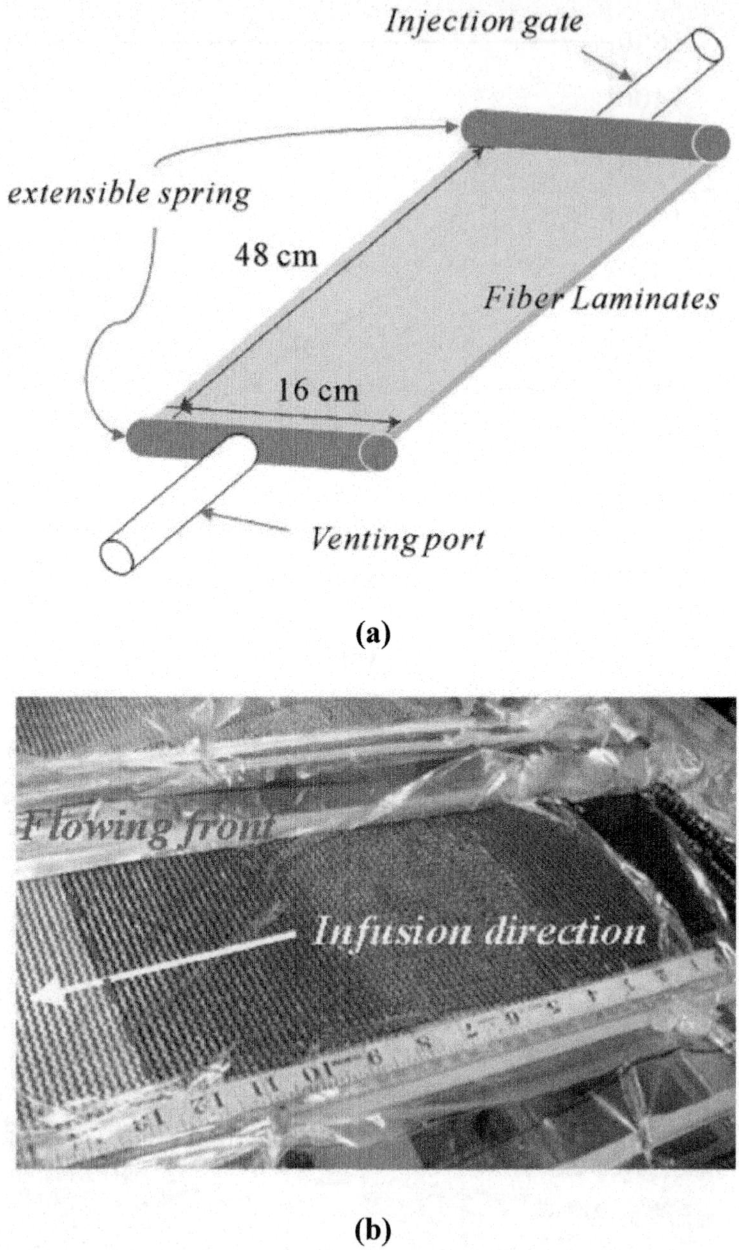

Figure 2: In-plane permeability experiment.

Table 1: Single-type experimental laminates

Designation	Weave Type	Lay-up numbers
M-300*	Strand Chopped Mats [isotropy]	1,2,4,8,12,16
M-450	Strand Chopped Mats [isotropy]	3,7
R-800	Woven Roving [0°/90°]	1,2,4,8,12,16
LT-800	Bi-axial Fabrics [0°/90°]	3,7
LT-1150	Bi-axial Fabrics [0°/90°]	3,7
DBLT-1900	Quad-axial Fabrics [-45°/0°/45°/90°]	3,7

*The number of each fiber weave is the area density (g/m^2).

The black and gray bars respectively represented the permeability values on the top and bottom surfaces. Measured permeability on the top surfaces was larger than that of the bottom surfaces because the distribution medium arranged on the top surface of fiber laminates and experimental permeability decreased as the same types of fiber layers increased (**Figure 4**(a)). **Figure 4**(b) shows experimental results of 7 fibrous layers in one assembly and the permeability decreased as the area-density of various fibrous weaves increased.

PERMEABILITY PREDICTION AND VERIFICATION

Permeability Prediction of Fiber Laminates with Single-Type Weave

The aforementioned steps derive the permeability of fiber laminates but the approach is time-consuming and inefficient.

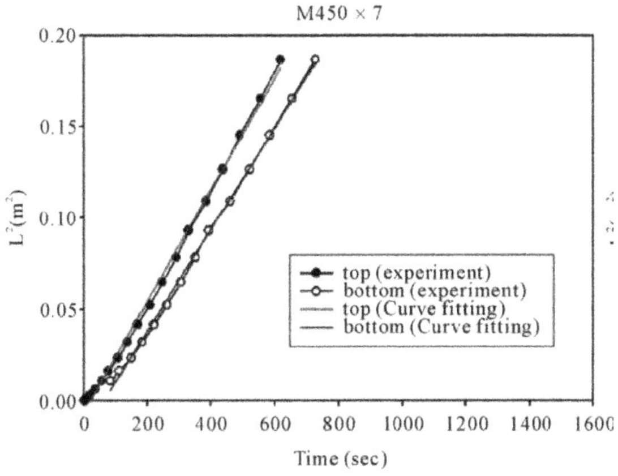

(a)

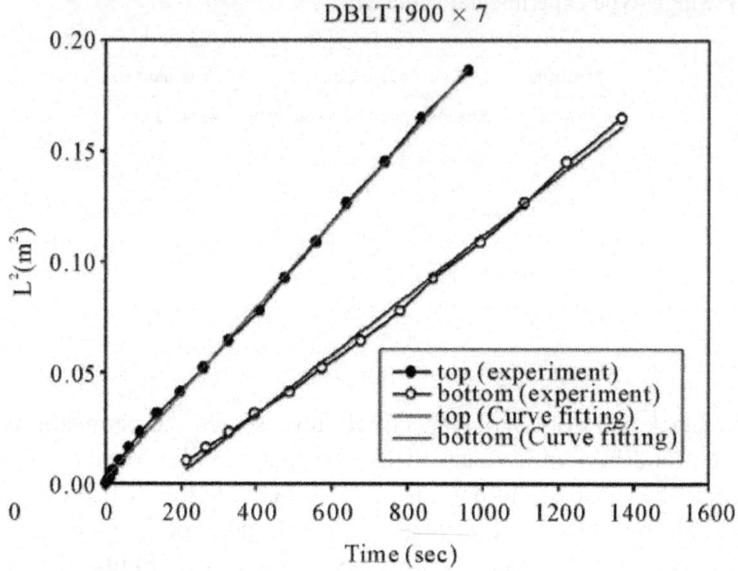

(b)

Figure 3: Flowing process of (a) M450 × 7 and (b) DBLT1900 × 7.

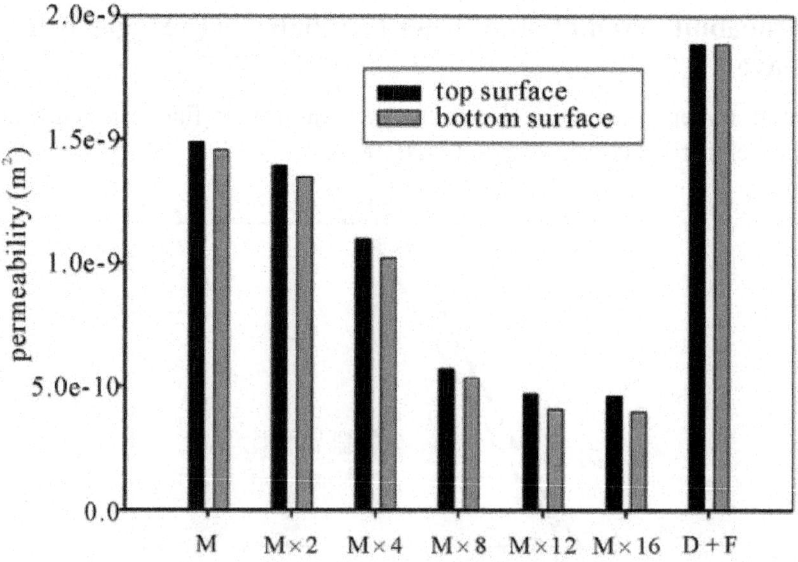

(a)

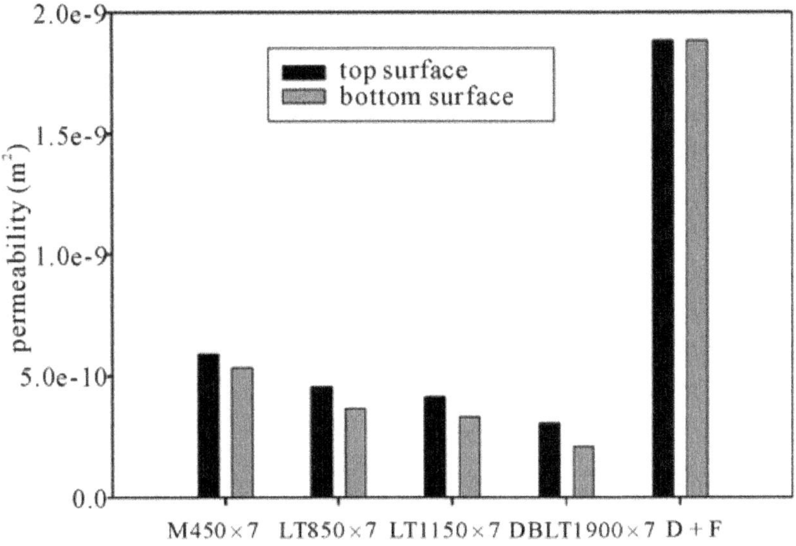

(b)

Figure 4: In-plane permeability of (a) Mat-300 and (b) various fibrous weaves.

Therefore, this research established a predictive method to derive the permeability of fiber laminates with no more operation of permeability experiments. Observing the experimental results revealed the in-plane permeability dropped as the increase of layer numbers and/or area-density of fiber weaves and changing these parameters meant changing the porous space inside FRP laminates. The length and width of each experimental specimen was the same, and hence thickness porosity, S_f, was defined as Equation (3) to display the change of porous space in the thickness direction.

$$S_f = t_f \times \phi_f \tag{3}$$

where t_f and ϕ_f are the thickness and the porosity of fiber laminate respectively. The parameter, S_f, provided a macroscopic viewpoint to discuss the resin flowing behavior and therefore the fiber diameter of weave was not considered herein to discuss permeability and flowing process [19].

The experimental data of top surface were rearranged in terms of thickness porosity, S_f (horizontal axis) and permeability, K (vertical axis) as **Figure 5** showed and the dots were experimental results including top and bottom surfaces. The data significantly exhibited a decreasing trend because larger thickness porosity, S_f, meant more porosity space inside fiber laminates that resin should infuse into. Hence two regressive curves were made to describe the relationship between thickness porosity and permeability on top and bottom surface. Equations (4) and (5) showed permeability tended to zero, as the flowing front stagnated, when the thickness porosity was extremely large, as limitless infusion space inside fiber laminates, and it conformed to the physical concept.

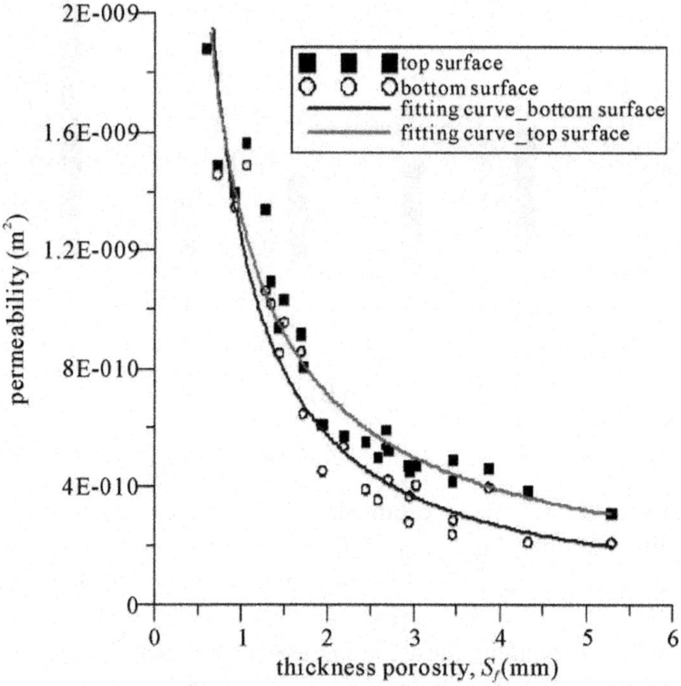

Figure 5. Experimental permeability of single-type fiber laminates.

$$\text{Top surface: } K = e^{-20.5} \times S_f^{\ -0.88} \tag{4}$$

$$\text{Bottom surface: } K = e^{-20.5} \cdot S_f^{\ -1.11} \tag{5}$$

Predictive Permeability Verification by Fiber Laminates with Multi-Type Weave

Figure 5 showed the permeability predictive curves of fiber laminates with single-type weave. However, the laminates applying in manufacturing large structures, such as yachts and wind turbine blades, are multi-type weave which includes chopped strand mats and different types of fiber weaves. **Table 2** displays the experimental laminates of multi-type weave as well as the measurements of thickness and porosity. The experimental permeability of each specimen derived from the in-plane flowing experiments and the predictive permeability obtained from the fitting curves showed in **Table 3**. The diversity of predictive and experimental permeability in **Table 3**almost below ±15% and the experiments with large diversity happened in thick laminates, such as (MD) × 4 + M and (MLT) × 5 + M, because the channel effect [8] easily occurred in the sides of thick laminates and affected the observation of flowing process. The permeability of single-type laminates with white dots and data of multi-type laminates with black dots on top and bottom surfaces displays in **Figure 6**(a) and (b) respectively. The permeability of multi-type laminates (black dots) superimposed on the data distribution of single-type laminates (white dots) and it meant the parameter of thickness porosity, S_f, was also suitable to predict the permeability of fiber laminates with multitype weave. The fitting curves were recalculated to get the new predictive equations as Equation (6), Equation (7), and **Figure 6** shows. Therefore the permeability of various fiber lay-up or sequences could be derived through the predictive equations without executing the time-consuming inplane permeability experiments, and the permeability was significant as running the simulation of infusion process.

$$\text{Top surface: } K = e^{-20.5} \cdot S_f^{\,-0.86} \tag{6}$$

$$\text{Bottom surface: } K = e^{-20.5} \cdot S_f^{\,-1.07} \tag{7}$$

SIMULATIONS AND EXPERIMENTS ON SHIP HULLS

Flowing Process Simulation with Various Infusion Scenarios

The predictive method for the in-plane permeability of fiber laminates was established and then was applied to manufacture a FRP ship hull. The target ship was designed for the Coast Guard Administration of Taiwan, and was measured 4.25 meters long and 1.66 meters wide. The designed laminate over the entire hull was (Mat-300 + DBLT-1900) × 2 + Mat-300 and its permeability was obtained from the predictive equations by only measuring the thickness and porosity of the designed laminate to input into the simulation software, RTM-Worx, to simulate the manufacturing process. Two injection strategies, the parallel scenario (**Figure** 7(a)) and the fish-bone scenario (**Figure** 7(b)), were used to analyze the simulated flowing pattern. The parallel strategy showed three parallel injection lines arranged in the longitudinal direction of the ship and the venting ports located around the hull side. The central injection gate above the keel was opened first, and the other two injection gates were triggered later as the flowing front passed through them about 30 cm. The fish-bone scenario showed one principal injection line along the keel, six minor injection lines perpendicular to the main line, and three injection gates on the principal injection line. Several simulations were executed to find the proper infusion arrangement including length of infusion lines, numbers and position of injection gates etc. for the target ship. Simulation before manufacturing is important and able to check whether the infusion time is less than the glue time and to ensure dry spots do not occur during the whole infusion process.

The simulated profile with the parallel strategy, **Figure 8**(a), showed a smooth front flowed in the whole infusion process and was beneficial to avoid dry spots as well as unsaturated impregnation. However, monitoring the flowing condition to decide the trigger time of each injection gate was a disadvantage of the parallel strategy, especially in manufacturing large components.

Table 2: Experimental laminates of multi-type weave

Laminate type	Abbreviation	Thickness (mm)	Porosity
(M300 + R800) × 1 + M300	(MR) × 1 + M	2.24	57.4%
(M300 + R800) × 2 + M300	(MR) × 2 + M	3.01	54.1%
(M300 + R800) × 3 + M300	(MR) × 3 + M	3.79	52.1%
(M300 + R800) × 4 + M300	(MR) × 4 + M	4.56	57.8%
(M300 + R800) × 5 + M300	(MR) × 5 + M	5.33	49.8%
(M300 + R800) × 6 + M300	(MR) × 6 + M	6.11	49.2%
(M450 + DBLT1900) × 1 + M450	(MD) × 1 + M	3.13	52.0%
(M450 + DBLT1900) × 2 + M450	(MD) × 2 + M	5.11	52.6%
(M450 + DBLT1900) × 4 + M450	(MD) × 4 + M	9.07	53.0%
(M450 + LT800) × 2 + M450	(MLT) × 2 + M	3.36	53.5%
(M450 + LT800) × 5 + M450	(MLT) × 5 + M	6.69	54.8%

Table 3: Predictive and experimental permeability on multi-type laminates

Laminate type abbreviation	Thickness porosity (mm)	surface	Predictive Permeability (m^2)	Experimental Permeability (m^2)	Diversity
(MR) × 1 + M	1.29	Top	1.11×10^{-9}	9.77×10^{-10}	−14.1%
		Bottom	1.03×10^{-9}	8.93×10^{-10}	−15.1%
(MR) × 2 + M	1.63	Top	8.90×10^{-10}	8.35×10^{-10}	−6.6%
		Bottom	7.78×10^{-10}	7.25×10^{-10}	−7.3%
(MR) × 3 + M	1.97	Top	7.31×10^{-10}	7.47×10^{-10}	2.2%
		Bottom	6.01×10^{-10}	6.38×10^{-10}	5.9%
(MR) × 4 + M	2.32	Top	6.18×10^{-10}	6.26×10^{-10}	1.4%
		Bottom	4.75×10^{-10}	4.97×10^{-10}	4.4%
(MR) × 5 + M	2.66	Top	5.37×10^{-10}	5.02×10^{-10}	−7.2%
		Bottom	3.86×10^{-10}	3.89×10^{-10}	0.9%
(MR) × 6 + M	3.01	Top	4.81×10^{-10}	4.63×10^{-10}	−3.8%
		Bottom	3.22×10^{-10}	3.56×10^{-10}	9.5%
(MD) × 1 + M	1.63	Top	8.92×10^{-10}	7.68×10^{-10}	−16.2%
		Bottom	7.80×10^{-10}	7.25×10^{-10}	−7.5%
(MD) × 2 + M	2.69	Top	5.32×10^{-10}	5.16×10^{-10}	−3.2%
		Bottom	3.79×10^{-10}	4.40×10^{-10}	13.8%
(MD) × 4 + M	4.81	Top	3.64×10^{-10}	3.62×10^{-10}	−0.7%
		Bottom	1.93×10^{-10}	2.99×10^{-10}	35.4%
(MLT) × 2 + M	1.80	Top	8.06×10^{-10}	6.54×10^{-10}	−23.3%
		Bottom	6.84×10^{-10}	5.61×10^{-10}	−21.9%
(MLT) × 5 + M	3.67	Top	4.13×10^{-10}	4.51×10^{-10}	8.5%
		Bottom	2.47×10^{-10}	3.67×10^{-10}	32.6%

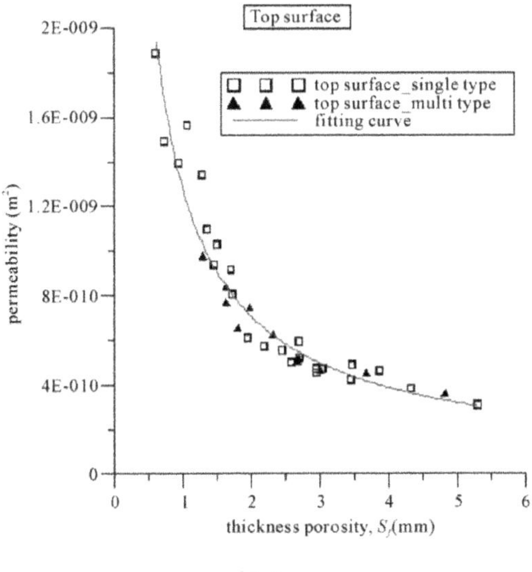

(a)

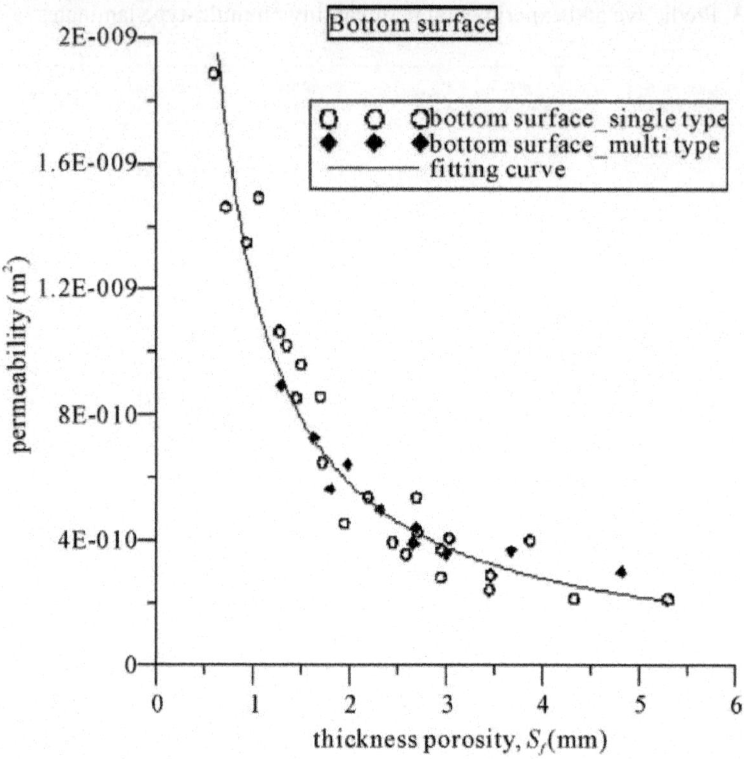

(b)

Figure 6: Permeability predictive curves on (a) top surface and (b) bottom surface.

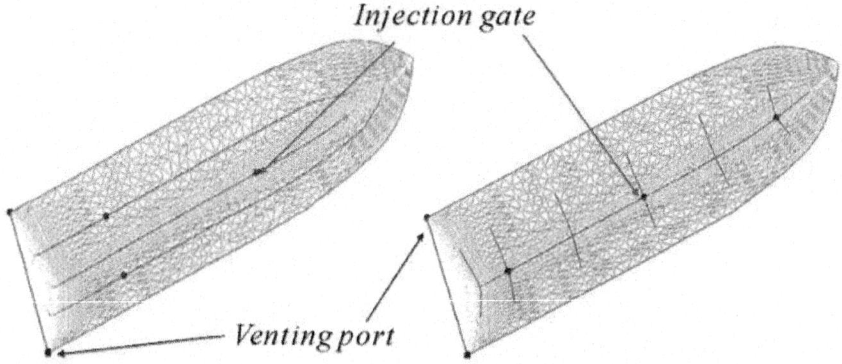

(a and b)

Figure 7: (a) Parallel and (b) fish-bone infusion scenario.

The fishbone infusion showed a sawtooth contour in the initial stage (**Figure** 9(a)) and finally had a smooth flowing front into the venting line. All injection gates triggered at the beginning of infusion was an efficient scenario without operating each injection gate separately. Exact simulation of the fishbone scenario before manufacturing was an important work to avoid infusion failures.

Infusion Measurements and Comparison with Simulations

Two manufacturing measurements of the ship hull with the parallel and fishbone scenario were completed to compare with the simulated results. Figures 8 and 9 showed the comparison contour between simulation and experiment with two different infusion strategies.

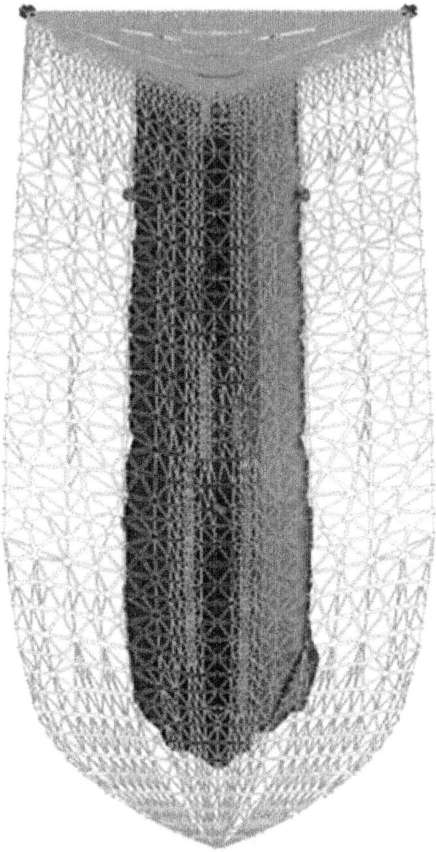

(a)

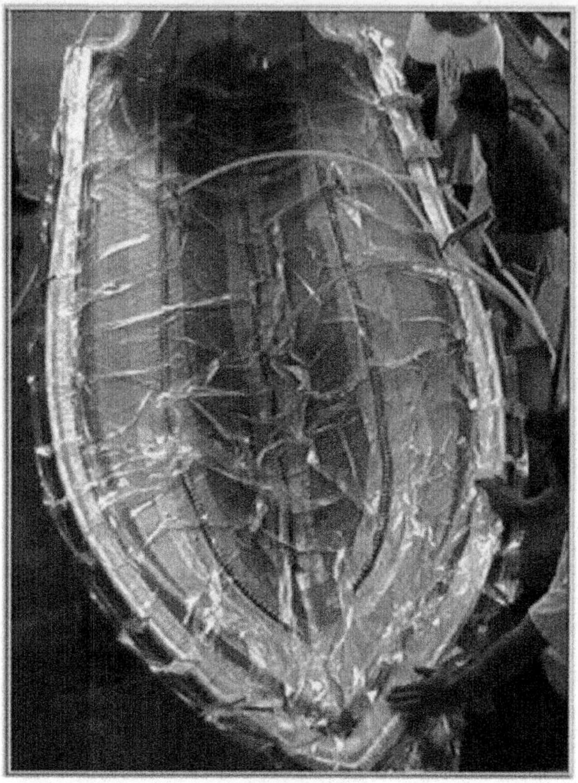

(b)

Figure 8: Simulation and experiment of parallel infusion scenario.

Both strategies displayed similar flowing shapes during the whole process. Total infusion time of two scenarios showed in **Table 4** and both experimental infusions were longer than the simulation time because of fiber overlap (**Figure 10**) near the keel to ensure structural safety. The small diversity explained the predictive method for the in-plane permeability of fiber laminates was validated and able to simulate the VARTM infusion process.

CONCLUSIONS

The permeability, K, reflecting the flowing ability of fiber laminates depended on the thickness porosity, S_f, in the macroscopic viewpoint. Although the compressibility effect during the flowing process was not discussed, the predictive method derived from experimental data had already included the effect.

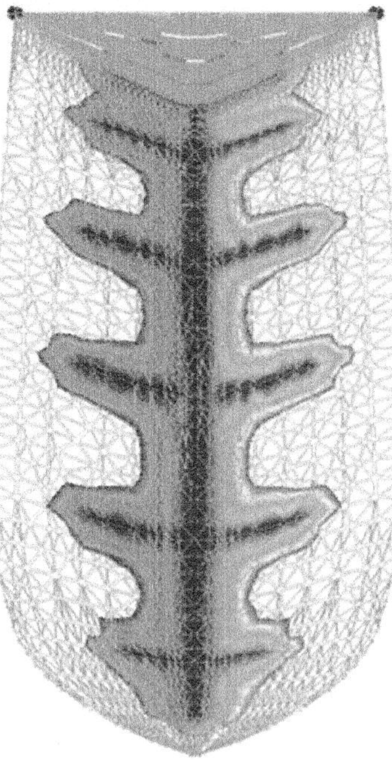

(a)

(b)

Figure 9: Simulation and experiment of fish-bone infusion scenario.

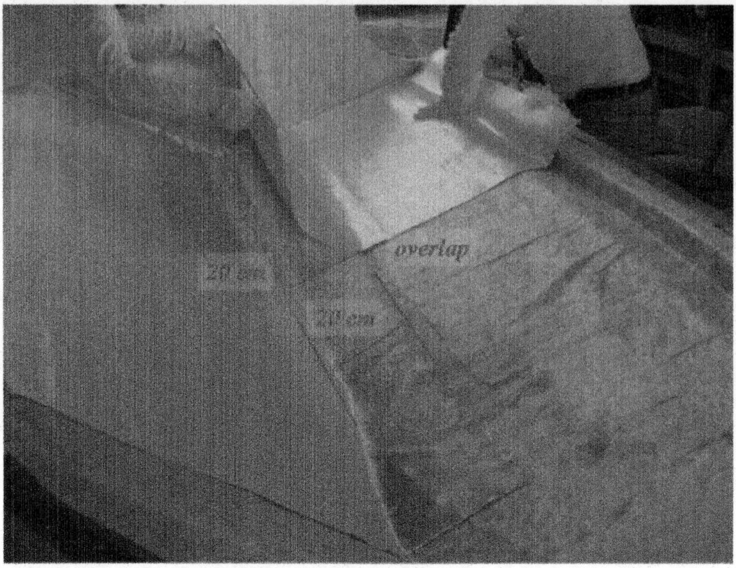

Figure 10: Overlap on the keel region.

Table 4: Experiments and simulations of ship-hull infusions

Infusion scenario	Experiment time (sec)	Simulation time (sec)	Diversity
Parallel	658	626	4.9%
Fish-bone	780	698	10.5%

Consequently, the predictive method by defining the parameter S_f was proposed to derive the in-plane permeability of fiber laminates under the VARTM process. The fine agreement between numerical simulations and experiments of a ship hull explained well applicability of the predictive method as manufacturing FRP products. Two infusion scenarios, parallel and fish-bone, had particular advantages. The experience of operating the software showed the parallel strategy was suitable for a slender and large structure, such as wind turbine blades, while the fish-bone strategy was applicable to manufacture structures without too large aspect ratio, such as a small boat. The predictive method for the in-plane permeability was established in the study and was beneficial to simulate the manufactures of FRP products under the VARTM process.

ACKNOWLEDGEMENTS

The authors would like to thank the National Science Council of the Republic of China, Taiwan for financially supporting this research. (NSC 94-2622-E-002-015-CC3).

REFERENCES

1. B. Pfund, "Resin Infusion in the US Marine Industry," Reinforced Plastics, Vol. 38, No. 1, 1994, pp. 32-34. doi:10.1016/0034-3617(94)90021-3
2. C. Williams, J. Summerscales and S. Grove, "Resin Infusion under Flexible Tooling (RIFT): A Review," Composite Part A-Applied Science and Manufacturing, Vol. 27, No. 7, 1996, pp. 517-524. doi:10.1016/1359-835X(96)00008-5
3. W. D. Brouwer, E. C. F. C. van Herpt and M. Labordus, "Vacuum Injection Moulding for Large Structural Applications," Composite Part A-Applied Science and Manufacturing, Vol. 34, No. 6, 2003, pp. 551-558. doi:10.1016/S1359-835X(03)00060-5
4. W. B. Young, K. Han, L. H. Fong and L. J. Lee, "Flow Simulation in Molds with Preplaced Fiber Mats," Polymer Composites, Vol. 12, No. 6, 1991, pp. 391-403.doi:10.1002/pc.750120604
5. L. J. Lee, W. B. Young and R. J. Lin, "Mold Filling and Curing Modeling of RTM and SCRIMP Processes," Composite Structures, Vol. 27, No. 1-2, 1994, pp. 109- 120.doi:10.1016/0263-8223(94)90072-8
6. R. V. Mohan, D. R. Shires, K. K. Tamma and N. D. Ngo, "Flow Channels/Fiber Impregnation Studies for the Process Modeling/ Analysis of Complex Engineering Structures Manufactured by Resin Transfer Molding," Polymer Composites, Vol. 19, No. 5, 1998, pp. 527-542. doi:10.1002/pc.10127
7. X. Sun, S. Li and L. J. Lee, "Molding Filling Analysis in Vacuum-Assisted Resin Transfer Molding, Part I: Scrimp Based on a High-Permeable Medium," Polymer Composites, Vol. 19, No. 6, 1998, pp. 807-817. doi:10.1002/pc.10155
8. J. Ni, S. J. Li, X. D. Sun and L. J. Lee, "Mold Filling Analysis in Vacuum-Assisted Resin Transfer Molding. Part II: SCRIMP Based on Grooves," Polymer Composites, Vol. 19, No. 6, 1998, pp. 818-829. doi:10.1002/pc.10156
9. A. Koorevaar, "Simulation of Liquid Injection Molding," Proceedings of 23rd SAMPE Europe Conference, Paris, 9-11 April 2002.

10. A. Koorevaar, "Fast, Accurate, Reliable 3D Reactive RTM Simulation," Proceedings of ISCM 2002 Conference, Flevoland, 30-31 May 2002.
11. T. J. Wang, C. H. Wu and L. J. Lee, "In-Plane Permeability Measurement and Analysis in Liquid Composite Molding," Polymer Composites, Vol. 15, No. 4, 1994, pp. 278-288.doi:10.1002/pc.750150406
12. C. Demaria, E. Ruiz and F. Trochu, "In-Plane Anisotropic Permeability Characterization of Deformed Woven Fabrics by Unidirectional Injection. Part I: Experimental Results," Polymer Composites, Vol. 28, No. 6, 2007, pp. 797-811. doi:10.1002/pc.20107
13. C. Demaria, E. Ruiz and F. Trochu, "In-Plane Anisotropic Permeability Characterization of Deformed Woven Fabrics by Unidirectional Injection. Part II: Prediction Model and Numerical Simulations," Polymer Composites, Vol. 28, No. 6, 2007, pp. 812-827.doi:10.1002/pc.20108
14. C. H. Wu, T. J. Wang and L. J. Lee, "Trans-Plane Permeability Measurement and Its Application in Liquid Composite Molding," Polymer Composites, Vol. 15, No. 4, 1994, pp. 289-298. doi:10.1002/pc.750150407
15. P. B. Nedanov and S. G. Advani, "A Method to Determine 3D Permeability of Fibrous Reinforcements," Journal of Composite Materials, Vol. 36, No. 2, 2002, pp. 241-254.doi:10.1177/0021998302036002462
16. V. M. A. Calado and S. G. Advani, "Effective Average Permeability of Multi-Layer Preforms in Resin Transfer Molding," Composites Science and Technology, Vol. 56, No. 5, 1996, pp. 519-531. doi:10.1016/0266-3538(96)00037-1
17. R. Chen, C. Dong, Z. Liang, C. Zhang and B. Wang, "Flow Modeling and Simulation for Vacuum Assisted Resin Transfer Molding Process with the Equivalent Permeability Method," Polymer Composites, Vol. 25, No. 2, 2004, pp. 146-164. doi:10.1002/pc.20012
18. C. Dong, "An Equivalent Medium Method for the Vacuum Assisted Resin Transfer Molding Process Simulation," Journal of Composite Materials, Vol. 40, No. 13, 2006, pp. 1193-1213. doi:10.1177/0021998305057429
19. B. R. Gebart, "Permeability of Unidirectional Reinforcements for RTM," Journal of Composite Materials, Vol. 26, No. 8, 1992, pp. 1100-1133. doi:10.1177/002199839202600802
20. P. Simacek, V. Neacsu and S. G. Advani, "A Phenomenological Model for Fiber Tow Saturation of Dual Scale Fabrics in Liquid Composite Molding," Polymer Composites, Vol. 31, No. 11, 2010, pp. 1881-1889. doi:10.1002/pc.20982

21. S. S. Rahatekar and J. A. Roux, "Numerical Simulation of Pressure Variation and Resin Flow in Injection Pultrusion," Journal of Composite Materials, Vol. 37, No. 12, 2003, pp. 1067-1082. doi:10.1177/0021998303037012005
22. W. D. Carrier III, "Goodbye, Hazen; Hello, KozenyCarman," Journal of Geotechnical and Geoenvironmental Engineering, Vol. 129, No. 11, 2003, pp. 1054-1056.doi:10.1061/(ASCE)1090-0241(2003)129:11(1054)
23. K. T. Hsiao, J. W. Gillespie Jr., S. G. Advani and B. K. Fink, "Role of Vacuum Pressure and Port Locations on Flow Front Control for Liquid Composite Molding Process," Polymer Composites, Vol. 22, No. 5, 2001, pp. 660- 667. doi:10.1002/pc.10568
24. A. R. Nalla, M. Fuqua, J. Glancey and B. Lelievre, "A Multi-Segment Injection Line and Real-Time Adaptive, Model-Based Controller for Vacuum Assisted Resin Transfer Molding," Composite Part A—Applied Science and Manufacturing, Vol. 38, No. 3, 2007, pp.1058-1069. doi:10.1016/j.compositesa.2006.06.021
25. K. Han, S. Jiang, C. Zhang and B. Wang, "Flow Modeling and Simulation of SCRIMP for Composites Manufacturing," Composite Part A—Applied Science and Manufacturing, Vol. 31, No. 1, 2000, pp. 79-86. doi:10.1016/S1359-835X(99)00053-6
26. M. K. Kang, W. I. Lee and H. T. Hahn, "Analysis of Vacuum Bag Resin Transfer Molding Process," Composite Part A—Applied Science and Manufacturing, Vol. 32, No. 11, 2001, pp. 1553-1560. doi:10.1016/S1359-835X(01)00012-4

Chapter 7

THE FLEXIBLE INTEGRATION OF MACHINE OBJECTS WITHIN DISTRIBUTED MANUFACTURING SYSTEMS

J M Edwards and I A Coutts

MSI Research Institute, Department of Manufacturing Engineering, Loughborough University of Technology, Loughborough, Leicestershire, LE11 3TU, UK

ABSTRACT

Manufacturing processes are required to adapt to change as businesses respond to global competition. The paper describes a framework for building distributed manufacturing processes based on an integrating infrastructure. Through a decomposition based on application function, application interoperation and application interaction, proposals are made for structuring the integration software required to flexibly implement distributed systems which can evolve to support required change. The paper describes the architecture used in a proof of concept implementation of a machine vision process made up of distributed application objects. Support for change is demonstrated through comparison with a conventional distributed system which does not make use of the sewices of an integrating infrastructure. In describina the system architecture. reauirements for 'SOW CIM buildina blocks which can be 'dugged' into an integrdng infrastmcture are defined. this work together with a range of approaches to application interoperation proposed by researchers at the MSI Research Institute are being further investigated through an EPSRC funded project entitled 'Manufacturing Software Interoperability: Steps towards lnteroperating Distributed Objects'.

INTRODUCTION

Manufacturing machines are primarily complex pieces of mechanical engineering with associated elecmnic programmable conwollers. They are designed with their immediate functionality in mind as standalone processes capable of being bought 'off the shelf. As such, the main design priorifies are often software execution speed, reliable processing and mechanical accuracy. Unfortunately when viewed as a building block of an integrated manufacturing system these machines are often insular, inflexible, and do not enable modification as business requirements change. Much of their inflexibility stems from a lack of consideration given to the way such systems may be integrated. Advances in computing technology have enabled the use of computer based machines within industry to become increasingly viable.

Their application in future integrated manufacturing systems based on distributed technology implies the need for viewing these machines as application objects, or CIM building blocks, which provide manufacturing services and can be combined to achieve some integrated manufacturing process. In order that these objects can form part of a useful open integrated manufacturing system, where they can offer services to open client applications, the requirements for implementation within an integrated manufacturing system must be established. This paper describes a framework for structuring the software required to link client applications with the services provided by a machine vision application object. This framework has been used to build a proof of concept distributed system implemented across Sun workstations and PC platforms.

The framework, together with proposed implementation mechanisms, provides an integration methodology which supports systemized implementation and change, within an open distributed system. The proposals represent a position on the migration path from conventional systems to future manufacfuring systems based on distributed interoperating object technology [4] which will become available as we approach the 21st century (typically through the work of the Object Management Group (OMG) [I3]). Current work at the MSI Research Institute, supported by the Engineering and Physical Science Research Council (EPSRC), is investigating the migration from conventional to distributed systems through the use of Object Request Broker technology, together with the investigation of mechanisms for object interoperation. The framework urooosed in this Daoer is based on elements which address the following issues:

- The architecture of softt CIM building blocks which can be plugged into and removed from a software infrastructure which underpins a CIM system. This involves the separation of the following three issues:
    - manufacturing application functionality;
    - application interoperation functionality;
    - application interaction functionality;

- the provision of interaction mechanisms which involve the use of services provided by a layer of infrastructural software, and the interfacing of alien devices such that they become compliant with these integration services;
- The structuring of interoperation mechanisms which involve the creation of a virtual vision server and corresponding support for client applications, through a set of vision service functions. The virtual vision server then requires a mapping onto the real vision application object

The paper briefly describes an integrating infrastructure known as CIM-BIOSYS (CIM Building Integrated Open SYStems) which provides the infrastnkural software alluded to in the three points above. CIM-BIOSYS has been created by researchers at MSI [l] through a government funded research project. Proposals in the paper are made through identifying requirements for a soft integrated vision machine, implemented on CIM-BIOSYS. These proposals include the definition of the architectural elements of soft CIM building blocks. The paper continues by describing the implementation mechanisms used to build a proof of concept integrated machine vision system.

The implementation described is in line with intermediate or next generation open systems. Specialist facilities to enable interaction between open applications and the vision application object are needed because specialist vision hardware is needed which cannot run CIM-BIOSYS software. The generation of a vision alien device driver for CJM-BIOSYS is thus necessitated. The significance of this implementation is not only in the mechanisms described for interaction and interoperation but also because it offers a migration path from current solutions, i.e. discrete devices, to future fully open distributed applications.

These fully open systems would be based on resources that can support a common integrating infrastructure such as that proposed by OMG in its CORBA (Common Object Request Broker Architecture) specification [13] The architecture for integration includes mechanisms which will support the requirement of machine vision systems to adapt to change. These mechanisms are based on the following principles.

The provision of structure through:

- the use of an overall architectural framework
- the use of a layered decomposition (within the complete soft integrated vision machine) incorporating interfaces based on virtual machine abstractions [15];
- The use of structured software where additional unforeseen requirements can be implemented within existing C code templates [5];

- An implementation based on the use of an integrating infrastructure where the infrastructure has underlying management facilities which can support change [19].

A SOFT INTEGRATED MANUFACTURING SYSTEM BASED ON AN INTEGRATING INFRASTRUCTURE

Researchers at the MSI Research Institute have for some years recognized the requirement for a single consistent interface for applications requiring integration level flexibility. This flexibility could be provided through the control and management of interprocess interaction and information sharing, ensuring provision for controlled change. To address this requirement an integrating infrastructure and platform of services was derived through practical experience of solving contemporary integration problems. This infrastructure has evolved into CIMBIOSYS. In its current implementation CIM-BIOSYS consists of a number of functional blocks as shown in figure 1. The manufacturing functiondapplications shown in figure 1, are, in this context viewed as being those processes which perform some part of a distributed, yet integrated manufacturing operation, i.e. a cell controller or a scheduling application. The device drivers hidehater for the diversity of both functionality and implementation of system resources. Qpical examples of system resources include shop floor machines, proprietary databases and human operators. The need for an integrating infrastructure, such as CJMBIOSYS, is now widely accepted by the manufacturing systems integration research community [6. 161. By enabling the creation of 'open' software, the use of an integrating infrastructure can promote the more systematic generation and change of flexibly integrated systems.

This provides a means of dealing with the high levels of complexity found in most manufacturing organizations [l, 19,201. The integrating infrastructure deals specifically with issues of application interaction, not application interworkinglinteroperation. Inter-working of application objects is supported within OS1 layer 7 by the specification of a coherent set of communications functions termed 'application service elements' (ASEs which typically include file transfer, virtual terminal, job transfer etc [8, 141). In general, an application layer protocol comprises a combination of these ASEs. Within the classifications of OSI, an integrating infrastructure such as CIM-BIOSYS is positioned at a level similar to that of an ASE. The CIM-BIOSYS platform of services is directly equivalent to an ASE while its configuration and management facilities are analogous to the proposals for mechanisms to support Open Distributed Processing [SI. The CIM-BIOSYS communications drivers are loosely based on the OS1 model layers 1 to 6. ClM-BIOSYS is a tool for integration, in its role as an ASE within layer 7 it offers managed integration services to applications. It can be used in combination with other ASEs or to embrace existing ASEs. Most importantly, the additional configuration and management facilities which make up CIM-BIOSYS and underpin its integration services enable open systems interaction to be implemented in a soft integrated manner, i.e they can adapt to required change.

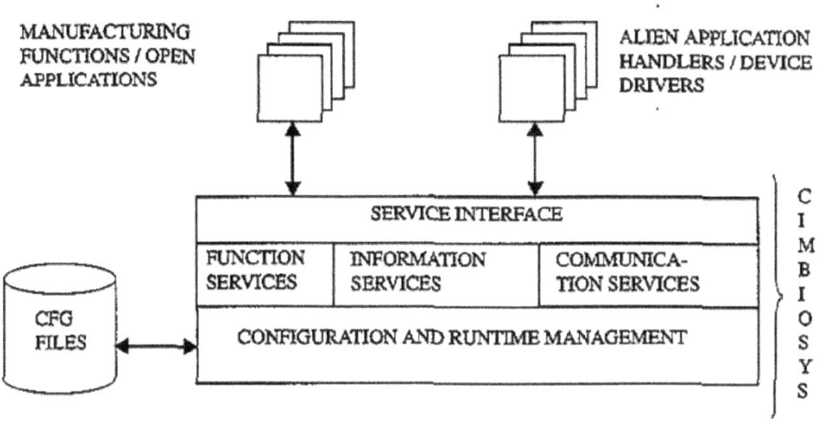

Figure 1: A functional view of CIM-BIOSYS.

It is proposed that further mechanisms are required to carry the 'soft' integration philosophy through to support application interoperation. These mechanisms should support the separation of application issues and interoperation issues which are considered to be conceptually separate [S] but are currently implemented with no explicit boundary. There is a need for an application service element (ASE) to support the interoperation of vision processing resonrceskervices and client manufacturing applications within an open distributed system.

BUILDING BLOCKS OF SOFT INTEGRATED MANUFACTURING SIRSTEMS

In order to generate a, soft integrated manufacturing system where application objects appear as discrete open applications (or soft CIM building blocks) within an integrated whole, two specific issues must be addressed:

- CIM-BIOSYS alien devices such as a remote vision processor must be handled in some way such that they appear to the CIM-BIOSYS software as compliant devices (i.e. provision of a CIM-BIOSYS alien device driver);

- the application objects must adhere to some predetermined messaging protocol such that they understand each other's messaging dialogue. If they are to be implemented as applications providing open interoperation they must be supported by specialist mechanisms (as proposed in this paper) and/or adhere to some recognized messaging standard.

Figure 2A shows the logical elements or objects within a machine vision system for electronic component inspection prior to placement on a printed circuit board. A detailed treatment of the information support aspects of the system are given in [3]. Figure 2B shows a mapping of the distributed vision system onto the CIM-BIOSYS integrating infrastructure. This figure illustrates how the complete logical solution for the information model driven inspection application can be implemented on CIM-BIOSYS. The 'operator interface and information management object' is implemented as a CIM-BIOSYS compliant open manufacturing application. The 'live component information generation' object is implemented as a vision application object on a CIMBIOSYS alien device. It is the mechanisms required within the relationship between these two objects which are the primary issues reported in this paper.

ARCHITECTURE FOR SOFT INTEGRATED MACHINE VISION SYSTEMS

The architecture proposed in this paper was developed and tested through the, implementation of the physical representation shown in figure 2B. The system comprised a vision client application, an information view provision application (not discussed in this paper) and the vision alien device driver running on any Sun workstation, distribution flexibility being provided by the CIM-BIOSYS integrating infrastiucture. This flexibility is provided through configuration facilities which provide a mapping between the logical elements of the distributed system and the physical whereabouts of the software objects in the system.

The vision server was implemented using a Matrox [9] vision processor installed in a PC. Figure 3 shows the layer by layer decomposition which make up the elements of the proposed architecture for a soft integrated vision machine. The decomposition between application object functionality, interaction issues and interoperation issues is deariy indicated, as are three discrete processes, i.e. The Vision Client Application, the Vision Alien Device Driver and the Remote Vision Service Provider.

THE OPEN INTERACTION OF APPLICATION OBJECTS

Structured and managed application interaction can be supported through compliance with the services -of an integrating infrastructure. To provide CIM-BIOSYS compliancy, enabling application interaction, applications should be able to invoke the CIM-BIOSYS interaction services and handle incoming CIM- BIOSYS data packets. Figure 3 shows the requirement for a 'CIM-BIOSYS Service Handling' module in a CIM-BIOSYS compliant manufacturing application.

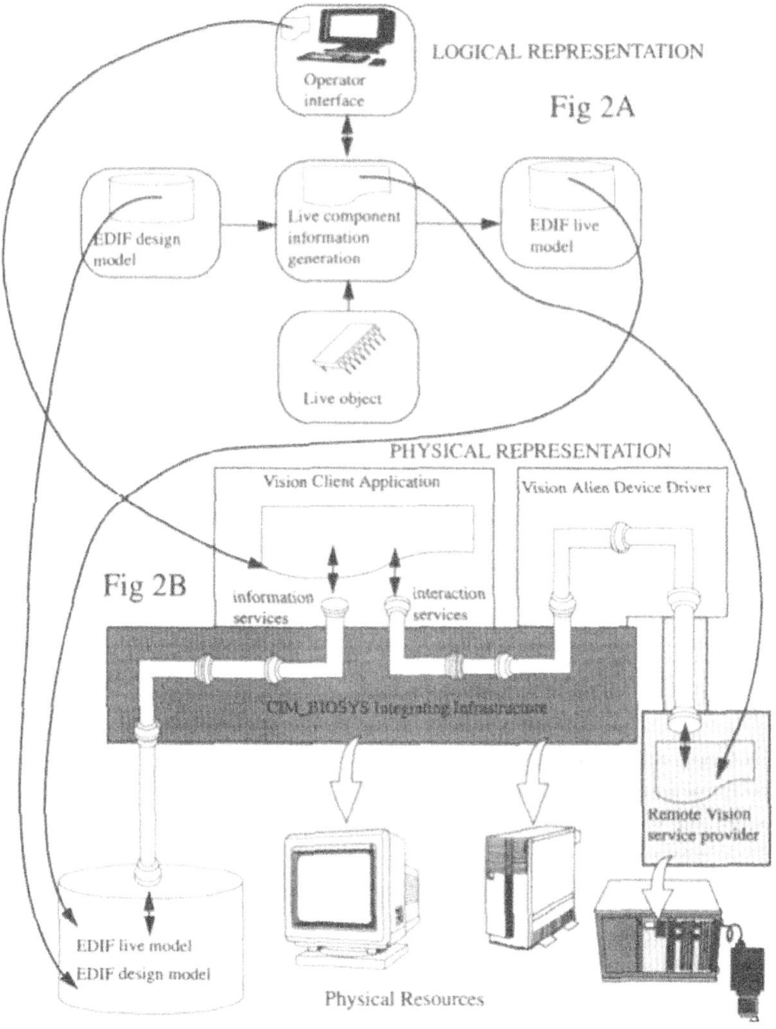

Figure 2: Mapping of the logical requirements onto the physical implementation.

Figure 3 also shows the additional requirement of an Alien Device Driver to provide CIM-BIOSYS compliancy for the Remote Vision Service Provider which contains the Vision Application Object. The Alien Device Driver must provide the functionality which enables the Remote Vision Service Provider to appear to Vision Client Application as another open application, i.e. the Remote Vision Service Provider will offer vision services on the CIM-BIOSYS integrating infrastructure.

To describe the functional requirements of the Vision Alien Device Driver, it is first necessary to detail the interaction aspects of the two processes which need to communicate—the Remote Vision Service Provider and the Open Vision Client Application. The Remote Vision Service Provider and the Alien Device Driver interact using socket based client/server interprocess communications (IPC) mechanisms [17], which exist within the 'Alien Device Server' (PC-NFS Sockets) and the 'Alien Device Client' (UNIX Sockets) within the Remote Vision Server and the Alien Device Driver shown in figure 3. The alien device client software runs on any general purpose Sun workstation on the network which is running CIM-BIOSYS. When the alien device driver is invoked, its initialization routine creates an IPC socket. It then builds and sends an 'initialisation' data packet containing a request for the status of the remote vision service provider. On receipt of a positive response from the server the alien device driver remains established to form its link between the vision server and open client applications on the CIM-BIOSYS infrastructure.

The Open Vision Client Application and the Vision Device Driver require a general purpose processing resource, they run on Sun workstations and interact through the use of the CIM-BIOSYS service interface. Implementation of the current version of CIM-BIOSYS also achieves IPC based on connectionless sockets [17]. The IPC transfers formatted data packets between CIM-BIOSYS and open applications. A detailed treatment of the implementation mechanisms used within CIM-BIOSYS is presented in [2]. The use of an integrating infrastructure for object interaction removes the necessity for multiple bespoke links between objects. Any client application with knowledge of how to use the vision services can apply to CIM-BIOSYS to establish a link with the remote vision server. The CIMBIOSYS alien vision device driver provides a single link between the vision server and any client. 'This approach helps to solve the complexity problems associated with distributed systems made up of many interacting objects. Objects are registered once with CIM-BIOSYS and all object interaction is managed through the use of CIMBIOSYS.

The Alien Vision Device Driver

The Alien Vision Device Driver has three principal functions which respond to external events, as shown in figure 4. These can be considered as three discrete modules as follows:

- the interface to the Remote Vision Service Provider through the Vision Service User, which responds to the reply messages sent from the remote system (this module is equivalent to the Alien Device Client Service User layer in figure 3);
- the interface to the CIM-BIOSYS integrating infrastructure through the CIM-BIOSYS Service User, which responds to request messages sent via CIM-BIOSYS from open applications requesting services from the Remote Vision Service Provider (this module is equivalent to the CIM-BIOSYS Service User layer in figure 3);
- the Alien Device Driver operator interface which provides manual control and monitoring of the driver for debug purposes, and responds to requests via a window interface on a Sun workstation.

Protocol conversion addresses the requirement to convert from CIM-BIOSYS compliant message packet format to the packet format used by the Remote Vision Server, and vice versa. This functionality is driven by, and is essentially part of, the two service user modules identified above. The event driven operation of the three modules which implement this functionality is controlled by the 'Notifier'.

The Notifier is a Sun tool [181 which provides a mechanism for distributing events to a number of functions within a process. The function of the Vision Service User (handler) is to field data packets from the Alien Device Server, apply error checking procedures, and to process the message. The principal process is to constmct a CIM- BIOSYS data packet using the complete vision server data packet as data, and sends it to the associated open client application via CIM-BIOSYS. The CIM-BIOSYS Service User (handler) is made up of a number of functions which are registered with CIMBIOSYS and are called in response to incoming CIMBIOSYS messages.

A typical function could terminate the association between the client application and the driver another could rem the status of the Remote Vision Server. This registering of device driver functions with CIM. BIOSYS is typical of the structured approach embodied in the use of such integration infrastructures. The approach helps solve the problems of building and modifying distributed software. A requirement for a new system component demanding the addition of a new alien device driver can be accommodated through the registration of additional functions with CIM-BIOSYS where these functions form the basis of the new driver. The knowledge of where and how to make these additions comes from the adoption of a reference architecture or framework. Once registered, CIM-BIOSYS, with its new driver provides open clients with managed access to the new remote object serverkystem device.

THE INTEROPERATION OF THE VISION SERVER AND THE VISION CLIENT

The previous section has discussed the interaction of system objects detailing how messages are securely transferred between objects in a structured system. This section describes the provision made for interpretation of the content of the messages which enables interoperation of the system objects. In this case interoperation is supported through the use of a virtual interface. The use of a virtual interface for interoperation enables a range of different variants of a class of manufacturing device to interoperate with client applications, such that all variants of the device appear to be uniform. The virtual interface is a mechanism which supports application interoperation, by using an abstract representation of specific functionality' provided by a real manufacturing machine [14]. Figure 3 shows the vision service user (or vision ASE) within the client device and the vision server (or virtual vision machine) within the alien device.

The Virtual Vision Machine (WM) is that portion of the Remote Vision Server that maps the services provided by the Vision Application Object on to the interaction mechanisms described in the previous section. The nature of the WM is then governed by the services provided by the Vision Application Object, and the interaction requirements of the remote device on which it is implemented. A vision machine can provide services which usually involve the processing of raw images, and can range from the provision of a set of application specific features of an object of interest, through classification decision information, to information relating to the understanding of the scene being viewed. The Vision Application Object in this work extracts application specific features, i.e. corner points of electronic component package legs. The full specification of the Vision Application Object requires the following interoperation with a Client Application: program invocation; information upload, and; information download. The Virtual Vision Machine (WM) operates on the decomposed data packet which is passed to it by the Alien Device Server (see figure 3).

This information comprises the package header and data buffer. The header contains a message identification field which controls how the WM deals with the message. Figure 3 shows the Vision Client Application has a layer dealing with the use of vision services. It is this area of functionality which engages in dialogue with the VVM. It must compose messages in accordance with the requirements of the WM and be capable of interpreting and dealing with the form of reply received from the VVM.

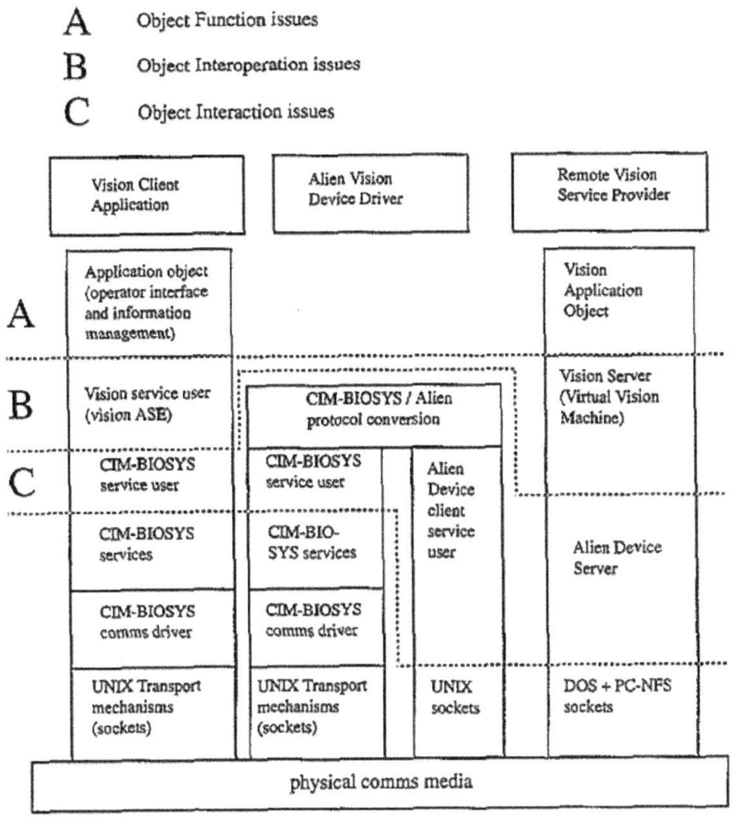

Figure 3: Elements within the Soft Integrated Vision System.

This work recognizes that in this case the problems of interoperation have been solved through specifying a particular set of vision service requests with specific semantics. The virtual interface provides flexibility in terms of replacing the vision application object. This philosophy demands the creation and adherence to international standards for manufacturing object interoperation. Although standards bodies have had some success in this area [lo] the philosophy suffers from an inherent lag in the system, i.e. when standards are agreed, technology has moved on.

Current work at MSI is tasked with providing a more flexible solution to object interoperation through systems mechanisms which aim to make available the semantics of object interaction at runtime me work described in this identifies the need for a separation between object function, interaction and interoperation issues, where the separation is clearly embodied in the implemented system. This separation be required; provides additional support for the modification of systems in line with required change.

A MODIFICATION TO PROVIDE AN ADDITIONAL VISION SERVICE

A modification to provide a new vision service has two distinct parts. Modification is required to the vision processing related issues, and also to the integration related issues to make the new service available to client applications. This paper only addresses the integration issues. The folloying subsection describes the modifications required to the integration elements within the complete Soft Intepted Machine Vision System. This work is measured relative to early experiments in integrating distributed applications without the use of an integrating infrastructure.

Modification within a soft integrated machine vision system

Figure 3 shows the elements of the Soft Integrated Machine Vision System. This figure helps to identify the extent of the required modification detailed in the following points:

- new facilities within the vision server, or virtual vision Machine in the Remote Vision Service Provider,
- new facilities within the Vision Service User, or Vision ASE in the Vision Client Application will be required, such that new open applications software can interoperate with the vision server;
- no modification is required within the Alien Device Server of the Remote Vision Service Provider;
- no modification is required within the Alien Vision Device Driver;
- no modification is required within the CIM-BIOSYS service user of the Vision Client Application.

The points above illustrate how all the modification is required within those elements of the system which implement the interoperation and application functions. No modification is required to the facilities provided to handle application interaction. This highlights the importance of the decomposition between interaction and interoperation. The new service requires the types of generic interoperation facilities identified as necessary for distributed machine vision. In this case modification within the VVM in the Remote Vision Provider, and the vision ASE within the Vision Client Application can be minimal.

Modification within a conventional distributed system

In a conventional dienuserver system it would be typical for all application functionality associated with communications issues to be contained within a single module of each application. This module would typically be structured as a set of functions implementing the interaction and interoperation requirements of the system, as suggested by MacKinnon [8]. The software to implement application interaction mechanisms within a distributed system are notoriously time consuming to develop and test.

A modification to implement a new client application which makes use of a new service provided by the server would require the implementation of the interaction mechanisms within the new client application. This requires knowledge of the interaction mechanisms used within the server software. It is this re-implementation of interaction facilities that is overcome by the use of the Alien Vision Device Driver within the framework of CIMBIOSYS.

The interoperation functions within the client and server of a conventional system would typically he implemented through Caset structures as is the case with the WM and Vision ASE of the soft integrated system. However, the identification of the generic functions within the VVM/ASE, together with the knowledge that the VVM/ASE is where new services are added are further elements of support that are missing in a conventional implementation.

RESULTS AND CONCLUSIONS

The changes made within the elements of the proposed integration architecture in order to provide the new service took approximately one third of the time taken to change the conventional communications software. The time consuming work on the conventional system was primarily for re-implementing socket based IPC from scratch, in the new Client application. Change to the software which implements the integration aspects of the machine vision system clearly highlights the advantages of implementing distributed processing using an integrating infrastructure. In particular, it shows the advantage of implementing the application interaction mechanisms once only for any number of open client applications.

The separation between interaction issues and issues of interoperation provides a modular decomposition where modification can be isolated within the mechanisms for interoperation. In the implementation described these mechanisms comprise the WASE combination. Further structure within the WM/ASE provides a modular template such that additional functionality can be &ily added to provide new vision services. The overall architecture provides a reference framework, and the interaction facilities are domain genetic and can be re-used. The VVM presents a consistent interoperation interface to new open client applications, while the ASE provides a set of ready built functions for inclusion within open applications to ease the use of the VVM services.

Heterogeneity is supported through the concept of virtual machines. Service specific software maps the VVM onto a particular Vision Application Object. Modification of this software enables a completely different Application Object, implemented using different vision hardware. The following points summarize the usefulness of the CIM-BIOSYS integrating infrastructure in underpinning the design, implementation and maintenance of the integration issues pertaining to distributed machine systems.

- *Soft Integrated Bnilding Blocks of Manufacturing Systems*: CIM-BIOSYS provides an underlying framework which structures the creation of manufacturing applications (or application objects) so that they become building blocks of flexibly integrated manufacturing systems. These building blocks can be 'plugged into' the CIM-BIOSYS infrastructure through 'registration' with the CIM-BIOSYS configuration information. (An example of a CIM-BIOSYS Open Manufacturing Application could comprise an executable C program, but any executable program capable of handling UNIX IPC is supported. It is this C program that is given a logical name and registered with CIM-BIOSYS.)
- *Reconfiguration*: The reconfiguration management mechanisms within CJM-BIOSYS provide distributed applications, and other manufacturing building blocks which interoperate with them, with a means of supporting long term flexibility and change. Reconfiguration takes place through modification of the CIM-BIOSYS configuration tables. This allows the association between logical applications and physical resources which make up the system to be modified without repercussion on the functionality of the distributed system.
- *Open services*: An Alien Vision Device Driver and its associated Remote Vision Service Provider provides a set of open vision services. These are application specific services offered at a level above that of the CIM-BIQSYS interaction services. Any open application registered with the infrastructure can access any of the vision services offered.
- *Heterogeneity*: The formal structure of CIM-BIOSYS enables the inclusion of the Alien Vision Device Driver. Alien device drivers can enable flexible integration of legacy building blocks and map device specific communication technology onto the CIM-BIOSYS interaction services. This provides a consistent interaction interface through which open applications can interoperate.

- *Overheads*: The use of CIM-BIOSYS implies an overhead in execution speed, size of the implemenfed system and time required to learn how to use the tool. Evaluation work which addresses the first two points (in relation to distributed manufacturing control systems) has been completed by researchers at MSI 1111. The runtime performance testing shows the overhead associated with the integrating infrastructure represents a small fraction of the overhead due to execution of the application systems themselves.

FURTHER WORK

As mentioned in the introduction to this paper work in the area of interoperating distributed object based systems is continuing at MSI with support from EPSRC. Object based technological innovation is available in many forms, typically many of the principles to support structured object distribution embodied in the CIM-BIOSYS integration infrastructure are now supported by commercially available first generation object request broker systems. All the major players in the computing arena have systems for object disxibution (typically IBM DSOM, Sun DOE). Some suppliers make proposals for support of interoperation, usually within specific domains, i.e. Open Doc. The current three year project seeks to utilize this emerging technology, initially in the area of manufacturing business process systems as opposed to the machine based processes described in this paper. The work is supported by both manufacturing companies (users) and manufacturing software vendors. The principal aim of the project is to develop a strategy for migration from current large granularity software products which do not map ideally onto the requirements of re-engineered manufacturing systems designed for optimum operation, to distributed systems comprising fully interoperating multi-vendor objects.

ACKNOWLEDGMENTS

The authors would like to acknowledge Jack Gascoigne and Paul Clements for their work within MSI on CIM-BIOSYS and systems interoperation. Also Professor Richard Weston who heads the MSI Research Institute.

REFERENCES

1. Clements P, Coutts I and Weston R 1993 A lifecycle support environment comprising open systems manufacturing modelling methods and the CIM-BIOSYS infrastructure tools Maple 1993 Symposium on Manufacturing Automation Pmgramming Language Envimnmenrs (Orrmva, Canadn)

2. Coutts I, Weston R H, Murgatroyd I S and Gascoigne J D 1992 Open Applications within Soft Integrated Manufacturing Systems Pmc. Int. Con3 Manx Automatwn (Hong Kong)
3. Edwards J, Clements P and Murgatroyde S 1993 Machine vision integration and inform;tion support - methods models and tools lnt. J. Comnuter Inteamted - Manufacturing 6 323-34
4. Hirsch B. Kuhlmann T. Marciniak Z and Masow C 1994 Intelligent Application Integration for Distributed Production European Workshop on Integrated Manufacturing System Engineering (Gretwble) pp 493-9
5. Kernigan B W and Ritchie D M 1988 The C Programming Language (Englewood Cliffs, NJ: Prentice-Hall)
6. Klittich M 1990 CIM-OSA integrating infrastructure-the operational basis for integrated manufacturing systems Int. J. Computer Integmfed Manufacturing 13 168-80
7. Kosanke K 1991 The European approach for an Open Architecture for CIM (CIM-OSAFESPRIT project 5288 AMICE Computing and Contra1 Engng I. (May)
8. MacKinnon D, McCmm W and Sheppard D 1990 An Introduction to Open Systems Interconnection (New York Computer Science Press, W H Freeman)
9. Catalogue (1055 St Regis Boulevard, Dorval, Quebec, Canada) Systems-Manufacturing Message Specification. Part 1: Service Definitions, Part 2 Protocol Specification Application and information support systems for planning and control in CIM Final ACME Report, grant no GWF 69192 (available from the MSI Research Institute at Loughborough University, UK)
10. SUN Microsystems 1990 Sunview Programmers Guide
11. Weston R H, Zhang P, Murgatroyd I S, Coutts I A and Hodgson A 1991 Soft Integrated Assembly Systems Proc. 4th World Con$ Robotics Research (Pittsburgh, USA)
12. Weston R H. Hodgson A, Coutts I A, Murgatroyd I S and Gascoigne J D 1990 Highly Extendable CIM Systems, Based on an Integration Platform Proc. CIMCON '90 Con& NIST(Gailhersburg, MA, USA)
13. SUN Microsvstems 1990 Network Proarammina Guide (~anu~l) Revision A
14. Object Management Group (OMG) 1991 The Common Object Request Broker: Architecture and Specification OMG Document number 91.12.1.-Revision 1.1. USA

15. Application and information support systems for planning and control in CIM Final ACME Report, grant no GWF 69192 (available from the MSI Research Institute at Loughborough University, UK)

Chapter 8

PARTICLE REDUCTION AT METAL DEPOSITION PROCESS IN WAFER FABRICATION

Faieza Abdul Aziz, Izham Hazizi Ahmad, Norzima Zulkifli and Rosnah Mohd. Yusuff

Universiti Putra Malaysia Malaysia

INTRODUCTION

Metal Deposition or metallization process is one of the processes in fabricating a wafer. A wafer is a thin slice of semiconductor material, such as a silicon crystal, used in the fabrication of integrated circuits and other micro-devices. Due to the nature on the process, it creates lot of particles, which would impact the next process if it were not removed. Particle deposition on the wafer surface can cause the circuit to malfunction; leading to a loss of yield. Cleaning process needs to be done after metal deposition process in order to remove the particles

Metal deposition, which has been constructed by several metal layers, allows the flow of current between interconnections. Each metal layers consist of three types of metal films such as Ion Metal Plasma Titanium (IMP Ti), Titanium Nitride (TiN) and Aluminum. The metal deposition started after the wafer has completed the "Tungsten Chemical Mechanical Polishing" (CMP) process.

Metal layers are deposited on the wafer to form conductive pathways. The most common metals include aluminium, nickel, chromium, gold, germanium, copper, silver, titanium, tungsten, platinum and tantalum. Selected metal alloy also may be used. The metal layer is shown in Figure 1 and the interconnection between metal layers is shown in Figure 2. The deposited metal(s) offers special functionality to the substrate. Typically, the metal aqueous solution is employed for the wet metal deposition process due to the consideration of its low cost and operation safety. Metallization is often accomplished with a vacuum deposition technique. The most common deposition processes include filament evaporation, electron- beam evaporation, flash evaporation, induction evaporation and sputtering. There are also two types of wet metal deposition processes – electrolytic and electro-less plating.

Sputtering and evaporation are well established as the two most important methods for the deposition of thin films. Although the earliest experiments with both of these deposition techniques can be traced to the same decade of the nineteenth century (Grove, 1852; Faraday, 1857), up until the late 1960s evaporation was clearly the preferred film-deposition technique, owing to its higher deposition rates and general applicability to all types of materials. Subsequently, the popularity of sputter deposition grew rapidly because of the need to fabricate thin films with good uniformity and good adhesion to the substrate surface (demand driven by the microelectronics industry) as well as the introduction of radiofrequency (RF) and magnetron sputtering variants.

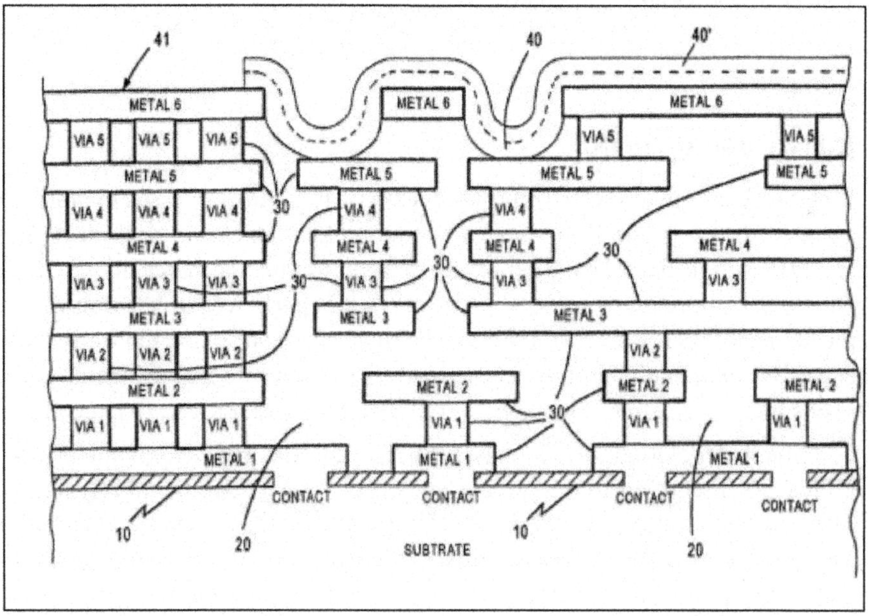

Figure 1: The Metal layers

In this chapter, a thorough investigation was carried out to improve shut down event problem at Metal Deposition process during wafer fabrication. Particle contamination on wafer surface can cause the circuit to malfunction and leading to machine shut down. Data of shutdown event versus sputter target life showed that the rate of machine shutdown increased by the increment of sputter target life. The sputter target life was further investigated to determine the appropriate sputter target life to be used in order to avoid particles generation during metal deposition process.

Figure 2: The interconnection between metal layers

PARTICLES

Particles can be defined as "suspension of solid or liquid mass in air". Particles can originate from a variety of sources and possess a range of morphological, chemical, physical and thermodynamic properties. The particles could be combustion generated, photo-chemically produced, salt particles from sea spray or even soil-like particles from re-suspended dust. Particles may be liquid; solid or could even be a solid core surrounded by liquid. Particles are represented by a broad class of chemically and physically diverse substances. Particles can be described by size, formation mechanism, origin, chemical composition, atmospheric behavior and method of measurement. The concentration of particles in the air varies across space and time, and is related to the source of the particles and the transformations that occur in the atmosphere. Some of the more generalized characterization of particles is:

- Primary and secondary particles: A primary particle is a particle introduced into the air in solid or liquid form, while a secondary particle is formed in the air by gas-to-particle conversion of oxidation products of emitted precursors.
- Particle characterization as per size: Particle can be classified into discrete size categories spanning several orders of magnitude, with inhalable particles falling into the following general size fractions- PM_{10} (equal to and less than 10 micrometre (μm) in aerodynamic diameter), $PM_{2.5\text{-}10}$ (greater than 2.5 μm but equal to or less than 10 μm), $PM_{2.5}$ (2.5 μm or less), and ultra-fine (less than 0.1 μm).

- Particle characterization depending on requirements of study: Some of the particle components/ parameters of interest to health, ecological, or radiative effects; for source apportionment studies; or for air quality modeling evaluation studies are particle number, particle surface area, particle size distribution, particle mass, particle refractory index (real and imaginary), particle density and particle size change with density, ionic composition (sulphate, nitrate, ammonium), chemical composition, proportion of organic and elemental carbon, presence of transition metals crustal elements and bioaerosols

Particle Contamination

Particle contamination can be defined as the act or process of contaminating by particulates. Particle contamination is problematic for many industries. They can appear unexpectedly mixed in solids, liquids and gases. Particles can be from many sources i.e.- metals, biological (skin, hair etc), polymers, building dusts etc. They all have different characteristics and properties such as shape, size and chemistry, which assist in identification. Scanning Electron Microscope (SEM) and Energy-dispersive X-ray spectroscopy (EDX) coupled with optical microscopy provides a powerful machine for unambiguously identifying such particles. The technique is frequently coupled with Fourier transform infrared spectroscopy (FTIR) when identifying the source of organic contamination (Stephen, 2010).

Particle Contamination in Semiconductor

Deposition of aerosol particles on semiconductor wafers is a serious problem in the manufacturing of integrated circuits. Particle deposition on the wafer surface can cause the circuit to malfunction, leading to a loss of yield. With the circuit feature approaching 1μm in size of one-megabit memory chips, particle control is becoming increasingly more important (Benjamin et al., 1987). Particle contamination during vacuum processing also has a significant impact in Very Large Scale Integration (VLSI) process yield (Martin, 1989) and has motivated most manufacturers to adopt particle control methods base on sampling inspection.

According to Bates (2000), semiconductor memory chips are very sensitive to the particles because the circuitry is so small. In a typical clean room manufacturing environment, particles are deposited on the wafer surface by sedimentation, diffusion, and/ or electrostatic attraction. Sedimentation usually occurs for large particles, particularly those larger than 1 μm in diameter, whereas diffusion occurs for small particles below 0.1 μm in diameter.

In the intermediate size range, both sedimentation and diffusion may occur and must be considered. When particles are electrically charged, enhanced deposition can take place. The rate of particle deposition on a wafer surface depends on both the size of the particle and their electrical charge. In addition, the deposition rate is also influenced by the airflow around the wafer, which in turn are affected by the size of the wafer, the airflow velocity, and the orientation of the wafer, with respect to the airflow. Although the mechanisms of particle deposition on semiconductor wafers are reasonably well understood and approximate calculations have been made (Cooper, 1986; Hamberg, 1985), no detailed quantitative calculation has been presented.

Particle Contamination in Wafer Processing

As the chip density increases and semiconductor devices shrink, the quality of fabrication becomes more crucial. The composition, structure, and stability of deposited films must be carefully controlled and the reduction of particulate contamination in particular becomes increasingly crucial as device sizes shrink and densities increase.

As the devices grow smaller, they become more sensitive to particulate contamination, and a contaminant particle size that was once considered acceptable may now be a fatal defect. Voids, dislocations, short circuits, or open circuits may be caused by the presence of particles during deposition or etching of thin films. Yield and performance reliability of microelectronic devices may be affected by the mentioned defects (Alfred, 2001). Often, the process gases will react and deposit material on other surfaces in the reactor besides the substrate. The walls of the processing chambers may be coated with various materials deposited during processing, and mechanical and thermal stresses may cause these materials to flake and become dislodged, generating contaminated particles. In processing steps that use plasma, many ions, electrons, radicals, and other chemical "fragments" are generated.

These may combine to form particles that eventually deposit on the substrate or on the walls of the reactor (Alfred, 2001). Particulate contamination also may be introduced by other sources, such as during wafer transfer operations and backstream contamination from the pumping system used to evacuate the processing chamber. In plasma processing, contaminated particles typically become trapped in the chamber, between plasma sheath adjacent to the wafer and plasma glow region. These particles pose a significant risk of contamination, particularly at the end of plasma processing, when the power that sustains the plasma is switched off. In many plasma-processing apparatuses, a focus ring is disposed above and at the circumference of the wafer to enhance uniformity of processing by controlling the flow of active plasma species to the wafer, such as during a plasma etch process. The focus ring, and the associated wafer clamping mechanism, tends to inhibit removal of the trapped particles by gas. Thus, there is a need to provide a reliable and inexpensive process to remove such particles from the wafer-processing chamber (Alfred, 2001).

Similarly, in chemical vapor deposition and etching, material tends to deposit on various parts of the apparatus, such as the susceptor, the showerhead, and the walls of the reactor, as the by-products of the process condenses and accumulates. Mechanical stresses may cause the deposited material to flake and become dislodged. These mechanical stresses are often caused by wafer transfer operations, but may also be caused by abrupt pressure changes induced by switching gas flow on and off and by turbulence in gas flow. Thus, process by-products at the end of the processing stage must be flushed from the chamber to prevent them from condensing and accumulating inside the chamber.

Typically, the flow of the processing gas is shut off at the end of a processing stage, whereupon the pressure in the chamber rapidly falls to zero as the vacuum pump continues to run. Idle purge may be used; in which purge gas is introduced into the chamber at intervals while no processing is taking place. Nonetheless, pressure spikes occur with the cycling of gas flow, causing disruption of particles, which may then contaminate the wafer surface. This limits the particle reduction benefits from the idle purge. A large portion of device defects is caused by particles disrupted by pressure change during wafer loading and moisture on the pre-processed wafer surface (Alfred, 2001).

Three types of particle contamination can be defined, which are under the deposited film as shown in Figure 3, in the deposited surface as shown in Figure 4 and deposited Film as shown in Figure 5. Particle under the deposited film will cause the surface of the wafer to become dirty. The particle may come from the previous process. Particle in the deposited surface will cause gas phase nucleation, leaks into the system, contamination in gas source/flow lines and sputter off walls. The particle may come from the gas phase nucleation, system leak or contaminated gas line. Particle on the deposited film will cause film build-up on the chamber walls. The source may come form the process chamber or from the wafer handling.

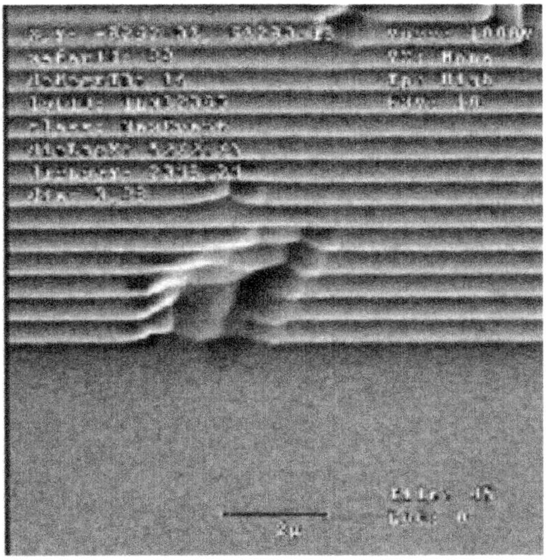

Figure 3: Particle under the deposited Film

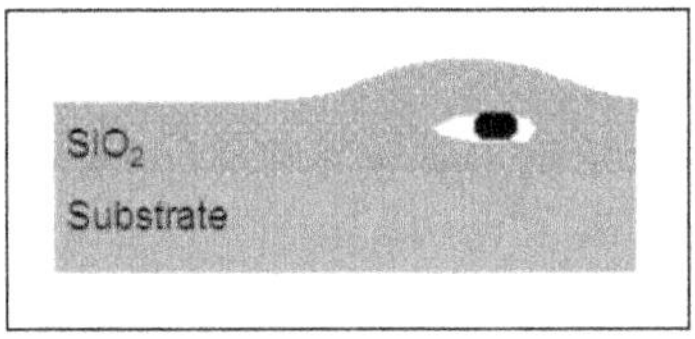

Figure 4: Particle in the deposited surface

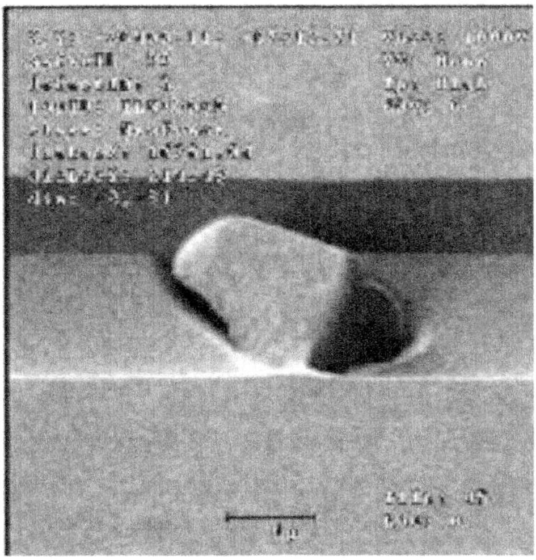

Figure 5: Particle on the deposited Film

Example of TiN particle transformation is shown in Figure 6. From this figure, the particle was dropped on the wafer's surface. The source of the particle may come either form previous process or from current process. After the deposition process done, the particle will be covered underneath the metal layer, which cause the damage of the interconnection.

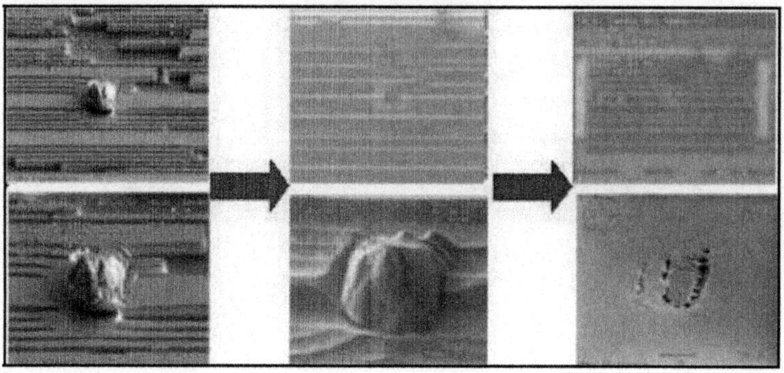

Figure 6: The transformation of TiN particle

The particles entrained in the load lock air volume by turbulence during pumping are either carried in through the handling of the wafer, generated within the camber from causes such as wear or residual from previous pumping and venting cycles. Particles are removed as they are drawn out during pumping or as they are carried out of the surface of the wafers. Additional particles may bind to the walls of the chamber or machining to tightly that they are agitated free by subsequent pump/ vent cycles (Peter Bordon, 1990).

An equilibrium background level is reached because the number of particles carried out by pumping and deposition on the wafer surface is proportional to the number of particles entrained into the gas volume. For example, if the number of particles entrained doubles, twice as many particles land on the wafer and twice as many flow out the pump line (Bordon, 1990). The effectiveness of these mechanisms has long been recognized. For example, it is a common practice to pump/ vent clean process chambers in high- current ion implanters and other process machines or to run getter wafers after a chamber has been contaminated.

Low levels of particulate contamination can be obtained in process gas systems by using careful system design, high-quality compatible materials, minimum dead legs and leak rates, careful start-up and operating procedures, etc. Low particle levels can also be obtained in gas cylinders through careful selection of cylinder materials, surface treatment and preparation, and through close attention to gas fill system design and operation (Hart, et al., 1994).

Particle levels in flowing gas systems may be steady or (as in machine vent lines) cyclic over time. In machine feed lines, the gas is usually well mixed and particles are uniformly distributed. However, particle levels in gas cylinders can vary by orders of magnitude over time due to such effects as liquid boiling, gravitational settling, and diffusion to internal surfaces. Such effects may also produce non-uniform particle distributions, including stratification, in gas cylinders (Hart, et al., 1995). Levels of suspended particles in filled cylinders can be measured with a high-pressure Optical Particle Counters (OPC). Data obtained directly from cylinders show that careful attention to quality can result in low cylinder particle concentrations.

Cylinder and bulk gases are frequently reduced in pressure with an automatic regulator before entering the flowing distribution system. Automatic regulators may produce increased particle levels (through regulator shedding, impurity nucleation, and condensational droplet formation) that are sometimes followed by system corrosion (Chowdhury, 1997) or suspended nonvolatile residue formation. Gases are therefore filtered after pressure reduction and before entering the distribution system. Ceramic, metal, or polymer membrane filters are selected for compatibility with the process gas. Such filters can produce a low particle level as well as a low degree of variability in contamination over time.

CNC data for particles as small as 0.003µm in O_2 and H_2 can also be obtained using an inert gas CNC with a special sample dilution device developed by Air Products (McDermott, 1997). These data showed that membrane filters can be used to produce high-cleanliness gases to 0.003µm in large-volume gas systems. Well-designed distribution systems should contribute a minimum of additional particulate contamination to the flowing gas. As the particle may impact the wafer quality, which results in wafer scrap, corrective and preventive action must be made immediately to stop the particle contamination from becoming catastrophic. Thus, a systematic problem solving method is needed to solve the issue.

Particle Failure

In this step, the types of particle failure were studied. Data from Daily Particle Qualification process was analyzed. TiN particle have highest standard deviation ~4.0 compared to Aluminium (Al) and Ion Metal Plasma Titanium (IMP Ti) as shown in Figure 7. This showed that TiN particle performed the most inconsistent compared to Al and Imp Ti particle in the Metal Deposition process. Data for each films particle qualification was obtained and then Pareto Analysis was made. From the Pareto Chart, TiN defect has the highest failure rate compare to other films as shown in Figure 8.

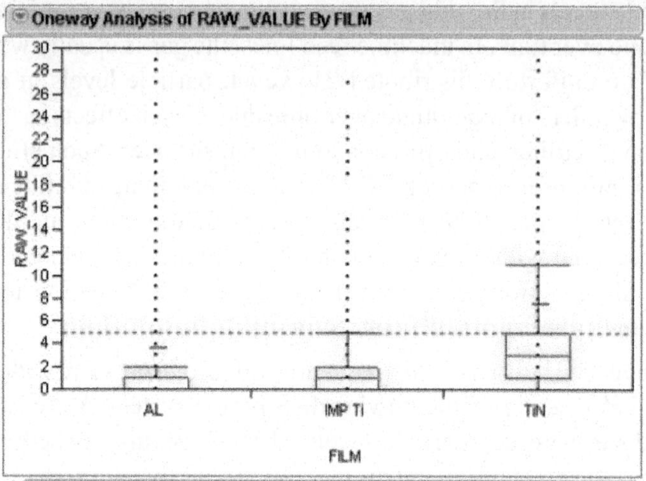

Quantiles

Level	Minimum	10%	25%	Median	75%	90%	Maximum
AL	0	0	0	0	1	3	29
IMP Ti	0	0	0	1	2	5	27
TiN	0	0	1	3	5	8	30

Means and Std Deviations

Level	Number	Mean	Std Dev	Std Err Mean	Lower 95%	Upper 95%
AL	2033	1.19085	2.49566	0.05535	1.0823	1.2994
IMP Ti	1192	1.89262	3.07781	0.08915	1.7177	2.0675
TiN	2070	3.64589	3.92741	0.08632	3.4766	3.8152

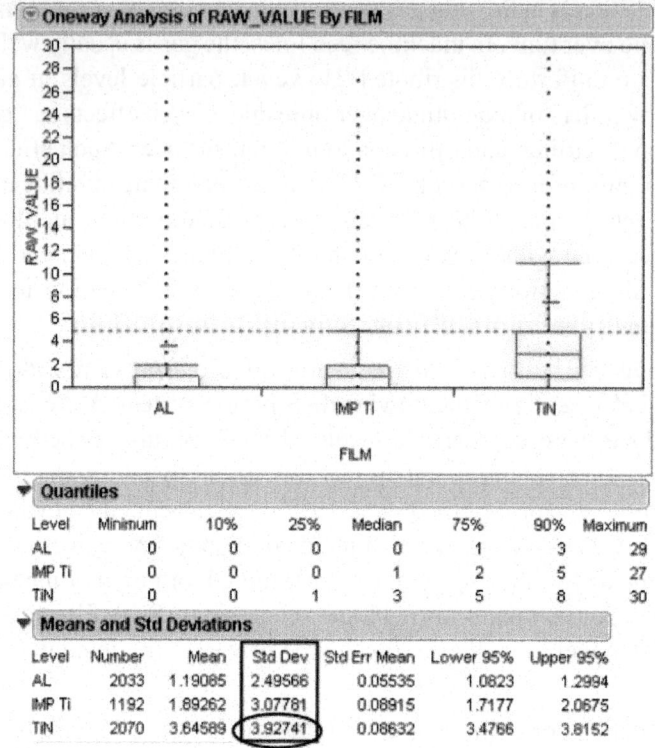

Figure 7: Comparison between Al, IMP Ti and TiN

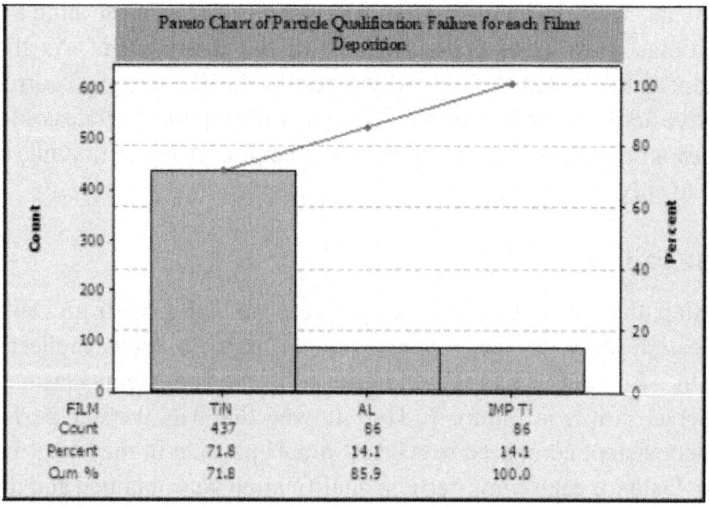

Figure 8: Pareto Chart of Particle Qualification Failure for each Films Deposition

Chamber Configuration and Wafer Processing Sequence

Chamber configuration for metal deposition machine is shown in Figure 9. From below chamber configuration, the wafers which inside the cassette are placed at cassette Load lock, which consist of Load lock A (LLA) and Load lock B (LLB). Wafer will pass through form Buffer camber to transfer chamber in Chamber A. Then, wafer will be cooled down in Chamber B. At the end of the process, wafers will be vented to Atmosphere condition in Load lock A or Load lock B, depending to which Load lock the wafers origin.

Wafers in a production pod will be pumped down to vacuum condition from atmospheric pressure in load lock A or load lock B, depending on where the lot is placed. Degas and notch alignment occurred in Chamber E and F. The deposition process begins with IMP Ti deposition in Chamber C. The wafers will move into the transfer chamber in Chamber A. Metal deposition will occur in Chambers 1, 2, 3 and 4. Depending on the application, multiple metal films can be stacked without breaking the vacuum. After the deposition process, wafer will be cooled down in Chamber B and vented back into the atmosphere in the production pod at Load lock A or B.

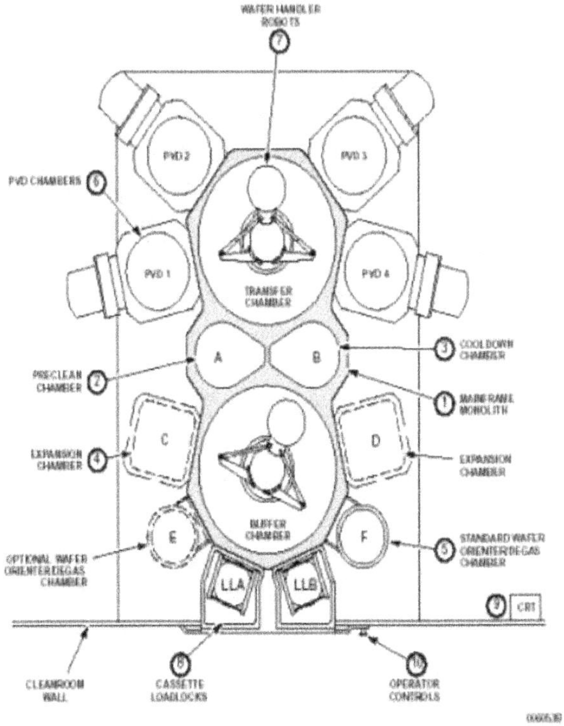

Figure 9: Chamber Configuration of the Metal Deposition Machine

Cause and Effect Analysis Diagram

A case study has been conducted in one wafer Fabrication Company. Downtime reduction at Metal Deposition process is being focused in this work. In average, three cases of Metal Deposition machines have been shut down every week. Shutdown criteria is based on more than "10-area count per wafer" or well known as "adders". The machine will be shut down if the post scan result shows particle increase more than 10 adders. Brainstorming session has been done with the team members and root causes have been identified and classified under six main sectors, which are machines, material, methods, measurements, environment and personnel. Figure 10 is the fish bone diagram for cause and effect analysis.

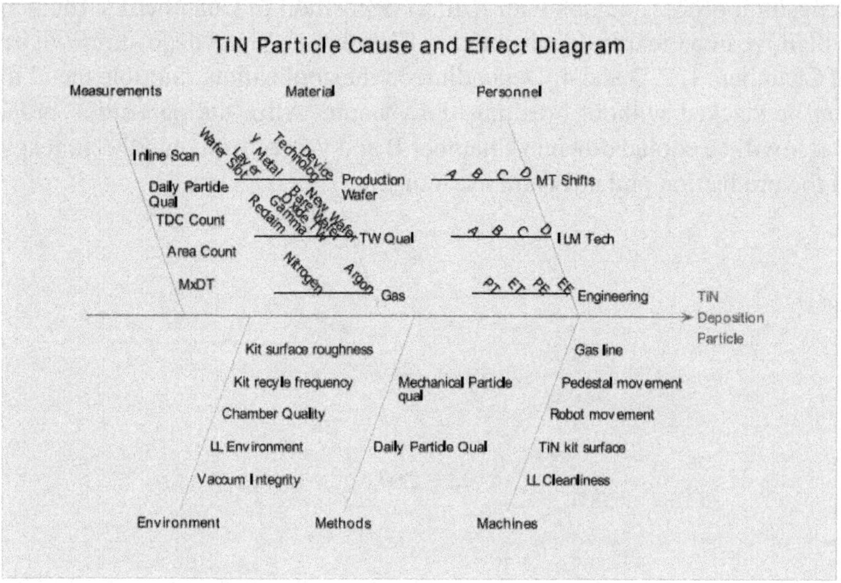

Figure 10: The fish bone diagram of cause and effect analysis

Measurement Systems

Particle measurement is performed by using the SP1 machine as shown in Figure 11. SP1 is the machine, which is measuring the particle by scanning and counting the existing particle on the wafer. Besides that, SP1 also is capable to show the wafer map, which can tell either the particles are clustered on the wafer, saturated, mild signature and others.

Percentage of Gage Repeatability and Reproducibility (GRnR) was done on SP1 in the production floor, which is used for particle scanning purpose.

Gage RnR study is conducted to determine the measurement system variability in term of Repeatability and Reproducibility. TiN particle count is our KPOV that causing machine shutdown by ILM due to metal deposition particle existence in the machine.

Based on the GRnR study, the result proves that the SP1 is capable to measure the TiN Dep particle counts cine the total GRnR is less than 30%. The result of GRnR study is shown in Figure 12.

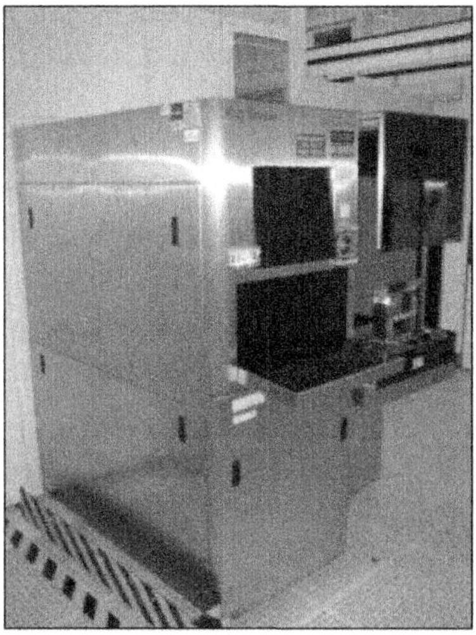

Figure 11: The SP1 machine which measure the particle

Gage R&R

Source	VarComp	%Contribution (of VarComp)
Total Gage R&R	0.67500	6.88
Repeatability	0.38333	3.91
Reproducibility	0.29167	2.97
Day	0.14722	1.50
Day*Slot	0.14444	1.47
Part-To-Part	9.13519	93.12
Total Variation	9.81019	100.00

Source	StdDev (SD)	Study Var (6 * SD)	%Study Var (%SV)
Total Gage R&R	0.82158	4.9295	26.23
Repeatability	0.61914	3.7148	19.77
Reproducibility	0.54006	3.2404	17.24
Day	0.38370	2.3022	12.25
Day*Slot	0.38006	2.2804	12.13
Part-To-Part	3.02245	18.1347	96.50
Total Variation	3.13212	18.7927	100.00

Number of Distinct Categories = 5

Figure 12: The result of GRnR study for SP1

In-line monitoring and systematic machine excursion monitoring

There are three methods of inspecting and measuring the particle, which are using production wafers through Systematic Machine Excursion Monitoring (STEM), production wafer that went to In-line monitoring process (ILM) flow and test wafers which is being used during machine qualification process.

In- line monitoring (ILM) is a process to detect any defect in real time. It is done in many ways, such as in line inspection, upon request from user, from production lots which go to ILM flow and also Systematic Machine Excursion Monitoring (STEM) lots. Machine-related defect excursions are controlled by systematically checking process machines. Production wafers are being used for STEM purpose. Each of Metal Deposition machine need to do STEM activity once every two days. STEM is a process where the lot which is already completely processed from one machine, will be held for ILM scan. The scan is done to check for any defects that may be caused by the processing machine at previous process. STEM will provide faster detection and containment of the defect excursion. All major process machines are monitored in a systematic manner.

For STEM activity, Manufacturing Technician will hold the lot for ILM Technician after run through the metal deposition machine. ILM Technician will scan four wafers/ machine using Complus or AIT machine. If the scan result shows particle signature and above the control limit (more than 10 counts), they will shut down the whole machine and the machine owner need to verify the shutdown prior to release the machine back to production.

For production wafers, there will be about 30% of the WIP will go to ILM inspection step. This is the random sampling in line scanning that has been designed to detect any defects along the process of fabricating the wafers from first process until end on the process. It has been designed in the process flow, where lots that are needed for this sampling will have ILM inspection flow compare to the other 70% of the lots that do not have ILM flow. Lots that have ILM flow will arrive at ILM inspection step after completing metal deposition process. ILM technician will scan the lot and if found particle and above the control limit (more than 10 counts), machine will be shut down and same verification need to be done prior to release the machine back to manufacturing. For qualification process, bare wafers or known as test wafers is used to check the machine's condition and performance.

Qualification process is done based on schedule. Basically, every metal deposition machines need to perform qualification process once everyday. This is to ensure the machine is fit to run and not causing any defect later. Qualification process is carried out by manufacturing technician using SP1. The qualification process is started with the pre particle measurement. Qualification wafer (bare wafer) will be selected and pre particle measurement is done using SP1. After premeasurement is completed, the wafer will go inside the machine and process chamber for machine and chamber qualification purpose. After the process is completed, the wafer is again brought to SP1 for post particle measurement.

The differences between pre particle value and post particle value will determine the machine and chamber's condition. For qualification process, the control limit is tightened to five count only. If particle is found more than five count, chamber will be shut down and pending verification from machine owner is needed prior to release the machine back to production. Example of the pre particle and post particle measurement is shown in Figure 13. In Figure 13, two wafer maps were shown, which are pre particle wafer map and post particle wafer map. In Pre Particle wafer map, two particle counts were detected as circled. In post particle wafer map, four particle counts were detected.

Two count were the existing particles and the other two were new particles, which were detected during post particle measurement. From Figure 13, the adders were two counts (post particle value- pre particle value).

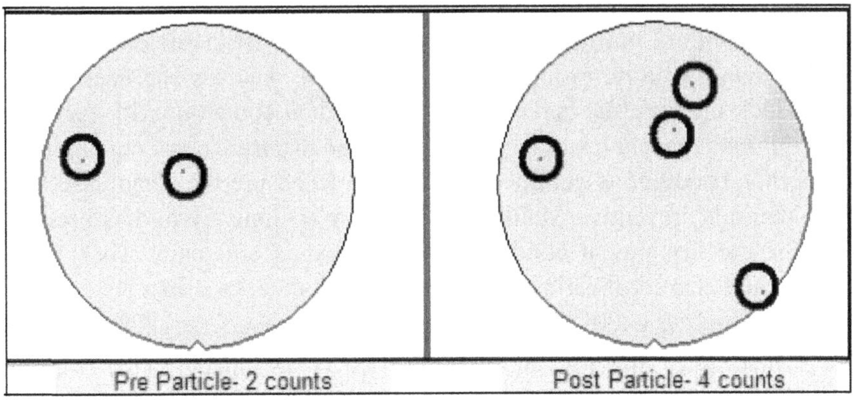

Figure 13: Pre Particle and Post Particle Wafer Map

Possible Root Cause

From Ishikawa Diagram, possible root causes will be screened out to get the actual causes. From actual causes, potential corrective and preventive actions will be determined and implemented.

Personnel

There are four shifts in the studied company, which are shift A, B, C and D. Manufacturing technician (MT) for every shift is responsible to perform the daily qualification job daily. Since there are four shifts running in the production, the level of experiences between shifts to shift differs. The level of experiences of the MT is very important since they need to perform the qualification process. Experienced MT will know and easily catch the particle issue inside the chamber by looking at the qualification result, but less experienced MT may take some times. Study has been done to check the Manufacturing Technician's efficiency. Their year of services and also certification were referred. The data were obtained from Human Resources Certification Record. Level of certification is from one to three. Level one is the minimum certification level, while level three is the maximum level of certification.

Study has been done to check the ILM Technician's efficiency. Their year of services and also certification were referred. The data were obtained from Human Resources Certification Record.

Process technician, process engineer, equipment technician and equipment engineer also play their roles during machine shutdown. When machine detected unwanted particle and need to be shutdown, process technician normally will follow the Out of Control Action Plan (OCAP) in order to release the machine back to manufacturing group as soon as possible. Process engineer also will take a look on the issue and do analysis and then come out with the release plan. Equipment technicians and engineers need to ensure proper maintenances job been carried out as per checklist. This is to ensure the cleanliness of the machine after Preventive Maintenance (PM) was done. Study has been done to check the Equipment and Process Technician's efficiency. Their year of services and also certification were referred. The data were also obtained from Human Resources Certification Record.

Beside MT, In-Line monitoring (ILM) technician also play big responsibility to determine the particle rate. It is important to have a proper scanning and analyzing of the STEM lot, so that the decision to shutdown the machine is base on real issue.

Material

For production wafers, different technologies will give different impact of the particle. This is mainly related to the process recipes, which different devices will have different process recipe, thus the deposition rate and thickness will be different from one device to others. Study has been done to see the relationship between particle issue and technologies.

From the shutdown event, list of lots that have been scanned was obtained. From the list, product technologies were segregated and the relationship between them with the particle is studied. Beside the technology, the metal layers also have impact to the particle issue since more metal layers means more times the lot will go to metal deposition process and the chances for expose to particle issue is more. Example for lot with four layer metal will go four times metal deposition process compare lot with five metal layers, will go five times metal deposition process. Study has been done to see the relationship between particle issue and the metal layers. From the same list from shutdown event, metal layers were obtained to see if there is any relationship between metal layers and particle. Test wafer also have some impact to the particle issue. For new test wafer, the performance is better compared to wafers that sent to rework and reused.

This is because the rework wafer normally will have remaining particle, which cannot be removed due to saturated at the surface of the wafer and needs stronger cleaning recipe to remove them. Brand new wafers normally will have a lot less particle. In this study, the incoming particle for 50 lots of new test wafers was measured using SP1 to get the potential incoming particle. From here, any existing particle from test wafers itself that may contaminate the process chamber later during qualification process can be seen.

Method and Measurement

Correct methods, which are used during both particle and mechanical qualification, were studied and observed. Judgment was made base on observation across all four shifts on the procedures during the qualification process.

The particle measurement is done based on In-line scan and during particle qualification process. For inline particle scanning, it is done after the lot has completed the metal deposition process. The job of in-line scanning is known as Systematic Machine Excursion Monitoring (STEM), which been done once in two days. Lot will be on hold for in line monitoring (ILM) scan.

Four wafers will be scanned for each machine to check for particle performance. The wafers will be scanned using Scanning Electron Microscope (SEM) machine. If particle signature exists, ILM personnel will notify the Metal Deposition machine owner to check for the machine's health. If the particle level exceeds the limit, which is more than 10 particle counts, the machine will be shut down and need to follow procedures in order to bring the machine back up to the production.

For production lots that go to ILM flow, the lot will be scanned for scratches and particle. If scratches or particle are found to exceed the limit, which is more than 10 particle counts, machine will be shutdown and need to follow the procedure as well. Wafers will be scanned using Complus or AIT machine also.

Qualification process is a process to check for the machine's performance, so that it always performs same as the baseline. One of the important factors in qualification is the particle performance. Particle value is measured based on the different value between pre particle measurement and post particle measurement. Differences of both values will determine the particle existence inside the machine. If particle count is more than five area count/ wafers, machine will be shutdown and need to be followed up by machine's owner before release back to production.

Machine

Machine is the main focus of the particle issue. This is due to the mechanical movement such as pedestal and robot movement inside the machine that can generate particle. Beside that, gas line also can create particle. Load lock cleanliness is also very important since this is the place where the lot is transferred into the machine from its base.

Load lock is a chamber that is used to interface a wafer between air pressure and the vacuum process chamber. According to Borden (Borden, 1988), Wu (Wu, et al., 1989) and Chen (Chen, et al., 1989), in the absence of a water aerosol, the dominant source of wafer contamination is the agitation of particles during the pumping (venting) of the entry (exit) load lock. In this study, 100 lots were selected to check the particle level in the cassette. Since wafers are inside the vacuum state inside the cassette, particle inside the cassette need to be measured. It was measured using mini- environment tester.

The cassettes were opened in Wafer Start room and the particle was measured for all the 100 lots. The particle count that obtained from the testing is captured. To study for particle during wafer handling and robot movement, mechanical qualification process was carried out. Before it was done, chamber was cleaned first to eliminate the potential source of particle coming from process chamber. One lot, which consists of five test wafers, was selected. Wafers inside the cassette were arranged in slot 5, 10, 15, 20 and 25. Pre particle measurements were obtained for all the five wafers.

The lot then was vented inside the load lock and also into the deposition module. Without running the deposition process, the lot was moved out back into the load lock and cassette. Post particle measurement was done to check for the adders. This cycle is repeated for 10 times for the entire machine and data is captured and analyzed. Particle in gas line also was focused in this study. Particle in gas line is measured by referring to the data that is obtained from the particle sensor.

The particle sensor is mounted at the gas line as shown in Figure 14. This is to ensure any particle in the gas line can be detected and the amount of particle entered to the process chamber can be monitored and recorded.

Sputter target also been studied to check the correlation between sputter target life and also shutdown. Example of sputter target that been used inside metal deposition machine is shown on Figure 15. The event of shutdown and the usage of target life is captured and analyzed.

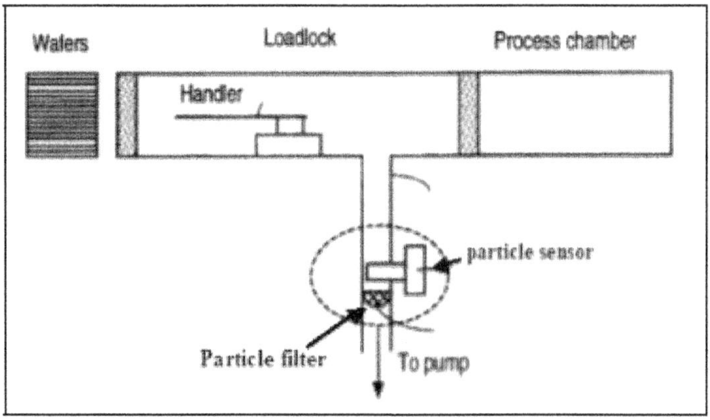

Figure 14: Particle Sensor that Mounted at Gas Line

Figure 15: Example of Sputter Target

Environment

The machine environment also has been studied to see any contribution to shutdown event. Load lock environment has been checked. Airborne particle measurement was conducted over metal deposition machines to collect the particle count. One hundred lots were prepared in this study. Wafers inside the cassette were arranged in slot 5, 10, 15, 20 and 25. Pre particle measurements were obtained for all the five wafers. The lot then was vented inside the load lock and left for five minutes. After five minutes, lot was moved out back from load lock move into cassette. Post particle measurement was done to check for the adders.

Results and Discussion

From the fish bone diagram, all the possible causes have been screened out and verified to find the actual true causes. From actual true causes, corrective actions and preventive actions will be defined, identified and will be implemented to eliminate the particle issue

Personnel

Verification have been made to people who directly working at metal deposition machine and related to the shutdown. Summary of the possible causes, which related to personnel, is shown in table 1.

Table 1: Summary of verify possible root causes related to personnel

Causes		Verification & Validation Process	Result	True/ False
Personnel	Manufacturing Technician (MT) in shift	1. To check the level of experiences of the MT 2. To check the capability of MT to perform qualification job correctly	1. Base on the study, all shift have dedicated MT > 2 years of experience to handle Metal Deposition tool 2. All the Manufacturing Technicians also capable to perform qualification job corectly	FALSE
	In Line Monitoring (ILM) Technician in Shift	To check the capability of ILM Technician to handle to scanning tool and catch the particle	Base on the study, all shift ILM Technician also having > 2 years of experience and capable to handle the scanning tool	FALSE
	Engineering Personnel (Process/ Equipment)	1. To check the capability of shift Equipment Technician (ET) and Equipment Engineer (EE) to perform Preventive Maintenance (PM) job efficiently. 2. To check the capability of Process Technician (PT) and Process Engineer (PE) to follow up on the ILM Shutdown issue to avoid re-shutdown	1. All shift ET/ PT and Engineers (PE/ EE) are well trained and have experiences > 3 years in average. 2. Equipment & Process team have their own checklist to be followed and verified by Section Head	FALSE

Manufacturing Technician (MT)

A validation and verification have been done to check the level of experiences of the MT and also the capability of MT to perform qualification job correctly. From the study, all shifts have dedicated MT more than two years of experience to handle Metal Deposition machine due to the criticality of metal deposition process. All the Manufacturing Technicians are also capable to perform qualification job correctly based on the checklist. The dedicated MT is summarized in table 2.

Table 2: Manufacturing Technicians in-charged of Metal Deposition Machine

Shift	Person in Charge	Date Join	Years of Experiences	Certification level
A	1. NEDUMARAN A/L MEGAWARNAM	2004	> 5 years	3
	2. LIYANA HANIM BINTI AKBAR	2006	> 3 years	3
	3. MUHAMAD TERMIZI BIN AHMAD TAJUDIN	2006	> 3 years	3
B	1. ROZI BIN MD HASSAN	2003	> 6 years	3
	2. NOOR JANNAH BINTI MAHADZIR	2007	> 2 years	3
	3. ANUAR BIN MAT ISA @ ABDUL AZIZ	2007	> 2 years	3
C	1. MOHD ASRIZAL BIN AHMAD	2003	> 6 years	3
	2. KASMINI BINTI TUKOL	2007	> 2 years	3
	3. MAIMUNAH BINTI HASHIM	2007	> 2 years	3
D	1. MOHD YUSRI BIN YUSOF	2007	> 2 years	3
	2. MOHD IZHAM BIN MOHD IZHAR	2004	> 5 years	3
	3. ZAIDA BINTI AHMAD	2007	> 2 years	3

Inline Monitoring (ILM) Technician

Based on the study, all shift ILM Technicians are also having more than two years of experience and capable to handle the scanning machine and captured defect images. The shift ILM Technician is summarized in table 3. Conclusion can be made that all the MT who handle the medal deposition machines are capable to perform qualification process and mistake that can lead to particle generation is almost zero.

Table 3: In-line monitoring (ILM) Shift Technicians

Shift	Person in Charge	Date Join	Years of Experiences	Certification level
A	REDZUAAN BIN ABDUL RAHIM	2002	>7 years	3
	BALAKRISNAN A/L A.MUNIANDI	2007	> 2 years	3
B	RUZAINI B. ADZHA	2006	> 3 years	3
	CHAREN A/L KHAN	2007	> 2 years	3
C	SUNTHARA MURTHI S/O RAMAN	2003	> 6 years	3
	ERUAN BIN ABU SEMAN	2007	> 2 years	3
D	VASANTHAN A/L VELOO	2007	> 2 years	3
	IBRAHIM BIN IDRIS	2005	> 4 years	3

Engineering Personnel (process/ equipment)

Verification and validation made to check the capability of shift Equipment Technician (ET) and Equipment Engineer (EE) to perform Preventive Maintenance (PM) job efficiently. Also validation made on the capability of Process Technician (PT) and Process Engineer (PE) to follow up on the ILM Shutdown issue to avoid re-shutdown due to incorrect qualification job done prior releasing machine during shutdown.

The summary of PT is shown in table 4 and summary of ET is shown in table 5. Based on the verification, all shift ET/ PT and Engineers (PE/ EE) are well trained and having experiences to perform their job efficiently. Equipment and Process team have their own checklist to be followed and verified by Section Head during performing PM activities and also releasing the machine from shutdown.

Table 4: Shift Process Technicians for Thin Film Metal Module

Shift	Process Technician	Date Join	Years of Experiences	Certification level
A	NOR AZELINA BINTI ISMAIL	2002	>7 years	3
	HARYANI BINTI ABDULLAH	2007	> 2 years	3
B	NORMALA BINTI NAPIAH	2006	> 3 years	3
	MOHD SYUKRI BIN CHE HASSAN	2007	> 2 years	3
C	KHAIRUL ANWAR BIN ABU BAKAR	2003	> 6 years	3
	CANITTHA A/P IEKIN	2007	> 2 years	3
D	PUTERI SURINAEDAYU BINTI MEGAT ISMAIL	2007	> 2 years	3
	NOR ADILA BINTI ABDUL RASHID	2005	> 4 years	3

Table 5: Shift Equipment Technicians for Thin Film Metal Module

Shift	Equipment Technician	Date Join	Years of Experiences	Certification level
A	MOHAMMAD RIDZAL BIN ABDULLAH	2002	>7 years	3
	MOHD KHADAFI BIN MAHAMOD	2007	> 2 years	3
B	AYUB BIN AHMAD	2006	> 3 years	3
	FRANCIS SELVAN A/L SINNAYAH	2007	> 2 years	3
C	MOHD ABDUL WAFI BIN AHMAD NADZIR	2003	> 6 years	3
	MOHD AZHUZAIRI BIN ABDULL AZIZ	2007	> 2 years	3
D	KHAIRUL HYFNI BIN NOORDIN	2007	> 2 years	3
	MOHD FAHMI BIN MOHD TAIB	2005	> 4 years	3

Conclusion can be made that PT who work at metal deposition process is capable to perform machine recovery as per procedure during the event of ILM shutdown. Particle generation during recovery or re-occurrence of shut down due to wrong recovery is zero. Conclusion also can be made that all ET who working with metal deposition machines are capable to perform preventive maintenance jobs effectively and mistake that can lead to particle generation is almost zero.

Material

Validation and verification were made in order to study the relationship between materials used in metal deposition process, with the shutdown rate related to particle, as shown in Table 6. The number of shutdown event (weekly) versus output (wafer move out from equipments) is shown in Figure 16. In average, three machines will be shutdown for every 27,900 wafer output from metal deposition process.

Relationship between shutdown event and output (weekly)

By using Minitab, correlation test between shutdown event and output has been conducted. The Pearson correlation between shutdown and output is 0.005, which means no correlations between both variables.

Regression analysis between shutdown event versus output was made using Minitab. The result of R square (R2) is zero, which means no relation between shutdown event and output.

Table 6: Summary of possible root causes related to materials used in metal deposition

Causes		Verification & Validation Process	Result	True/ False
Material	Production Wafers	To check the relationship between shutdown and moves	1 Correlations test between shutdown and moves had been done The Pearson correlation between shutdown and move is 0.005, which means no correlations between both variables 2 Regression analysis between shutdown versus Moves was made using Minitab as shown in Figure 4.3 From the result, the R square (R2) is zero, meaning no relation between shutdown and moves	FALSE
		To check the relationship between shutdown and moves by technologies	1 Pearson Correlation test was done between shutdown and each of the moves by technologies as shown in Figure 4.6 2 From the Pearson Correlation test, conclusion can be made that no correlation is exist between shutdown and technologies 3 Base on the regression analysis, no significant relationship between the technologies and the shutdown event The R2 also showed 0%, which meaning no relationship between the shutdown and technologies	FALSE
		Relationship Between Shutdown and Technologies (STEM Failed)	1 From the Pearson Correlation shown in Figure 4.8, conclusion can be made that no strong correlation between shutdown and technologies 2 Regression analysis was made between the shutdown and technologies as shown in Figure 4.9 and no strong evidence to conclude that the shutdown is influenced by the technology	FALSE
		Relationship Between Shutdown and Metal Layers (LM)	1 From Figure 4.11, no strong evidence to say that the shutdown is influenced by the four layer metal lots 2 From Regression analysis in Figure 4.12, also indicates no strong correlation between shutdown and layer metal	FALSE
	Test Wafers	Relationship Between Incoming Particle from New Test Wafer and Shutdown	All the 50 lots of new test wafer were passed the incoming particle screening From this study, conclusion can be made that the particle generation is not coming from the new test wafers that were used during qualification process.	FALSE

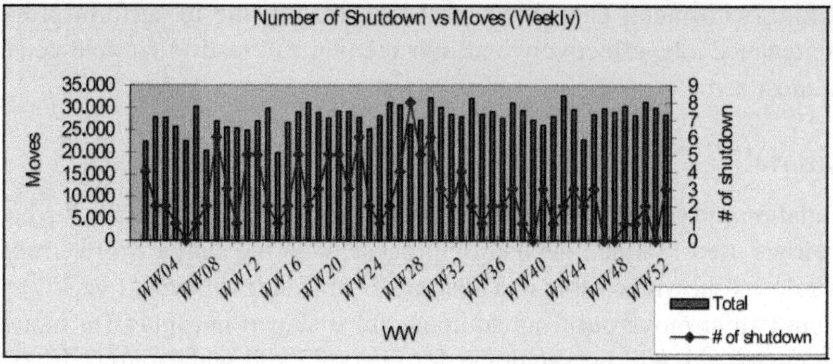

Figure 16: Number of Shutdown Event versus Machines Output (Weekly)

Relationship between Incoming particle (from new test wafer) and shutdown event

Incoming particle screening was performed for 50 lots of new test wafers. Histogram was generated as shown in Figure 17. From the results, the mean is 1.18 with standard deviation of 1.24. Out of 50 lots of new test wafer that have been measured, 23 lots resulted in zero count of incoming particle, 6 lots showed one count of incoming particle, 10 lots showed two count and 11 lots showed three counts of incoming particle. Since the specification for the incoming particle is five count, all the 50 lots of new test wafer passed the incoming particle screening. From this study, conclusion can be made that the particle generation is not caused by the new test wafers that were used during qualification process.

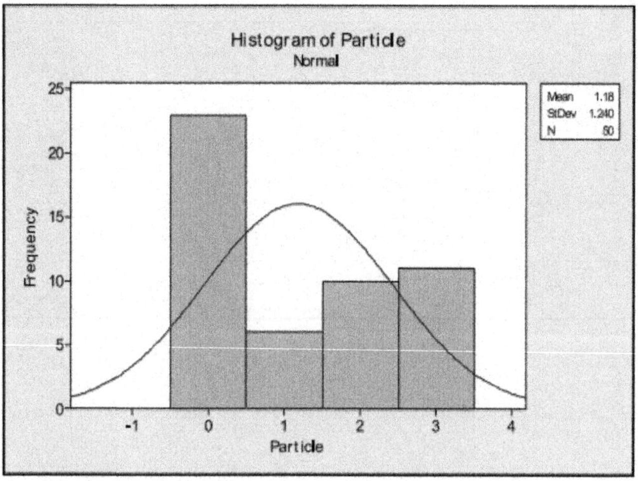

Figure 17: Histogram for Incoming Particle in New Test Wafer

Method and Measurement

Method of executing the qualification process were observed and summarized. Since the entire MT whose handle metal deposition machines were at level three, therefore they are considered as competent to perform the task efficiently. This is proved by the qualification data that is available in the spreadsheet and also in the CIM system. By looking at the shutdown trend, conclusion can be made that STEM is effective to detect particles that generated at Metal Deposition process.

Machines

The relationship between shutdown event and machine was studied and the summary of the result is shown in table 7.

Table 7: Summary of verify possible root causes related to Metal Deposition Machine

Causes		Verification & Validation Process	Result	True/ False
Machine	Vacuum Cassette	To check the particles that may enter a cassette from the cleanroom	Particle and mechanical qualification was done to check the particle existance. Result was clean and no particle was found during wafer loading from production pod to the loadlock	FALSE
	Wafer Handler/ Robot movement	To check the particle that may be created during wafer transfer from vacuum cassette to deposition module	Particle and mechanical qualification was done to check the particle existance. Found that particle was created during wafer loading from vacuum cassette to deposition module	TRUE
	Gas Line	To check the particle in the gas line	Base on the particle data which is obtained from the particle sensor, no partilce can escape through the particle filter	FALSE
	Sputter Target	To check the relationship between Sputter Target Life with ILM Shutdown	Base on the correlation analysis, there is relation between Sputter Target Life and ILM Shutdown	TRUE

Relationship between shutdown event and particle in vacuum cassette

Results of particle existence in vacuum cassette are shown in Figure 18. From the bar chart, 82 lots detected zero particle count, 13 lots showed particle with one count and five lots showed two counts of particle. Conclusion can be made that particle almost not exist and can be considered as negligible in vacuum cassette, as the production is running under clean room environment of class one category.

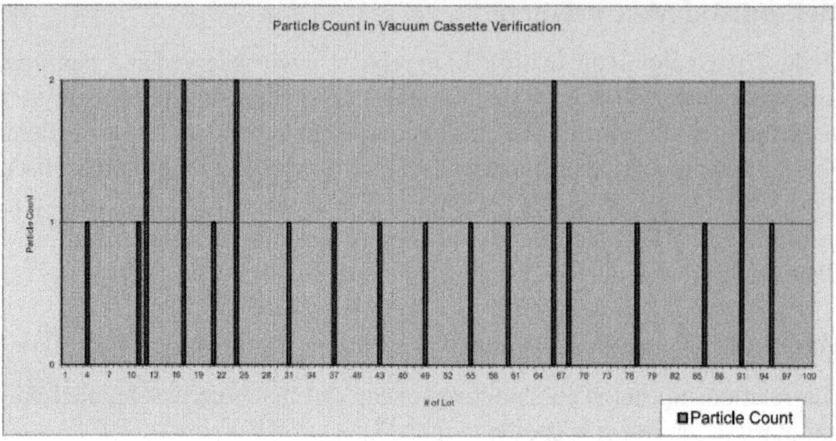

Figure 18: Bar Chart for Particle Existence in Vacuum Cassette

Relationship between shutdown event and particle generation during mechanical movement

Results for mechanical qualification process are tabulated in table 8. From the data, bar chart was generated as Figure 19. From the bar chart, each of the machines showed particle generation inside the load lock. From the data, conclusion can be made that particles can be generated during wafer handling and mechanical movement.

Table 8: Result for Mechanical Qualification Process for all Metal Deposition Machines

	Particle Count (adders) for mechanical Qualification Process for all Metal Depositon Tool									
Tool	1st	2nd	3rd	4th	5th	6th	7th	8th	9th	10th
Metal Dep 01	8	1	5	11	2	1	5	5	9	4
Metal Dep 02	12	3	8	2	2	4	12	5	9	8
Metal Dep 03	3	7	7	14	9	1	7	11	1	7
Metal Dep 04	15	1	7	2	10	2	5	8	2	9
Metal Dep 05	6	7	14	7	5	4	12	6	12	8
Metal Dep 06	3	2	1	3	5	13	9	3	7	11
Metal Dep 07	13	9	4	13	4	1	7	2	15	7
Metal Dep 08	5	3	9	4	2	7	9	3	1	5
Metal Dep 09	4	7	5	4	3	13	1	9	5	9

Figure 19: Particle Count for Mechanical Qualification Process

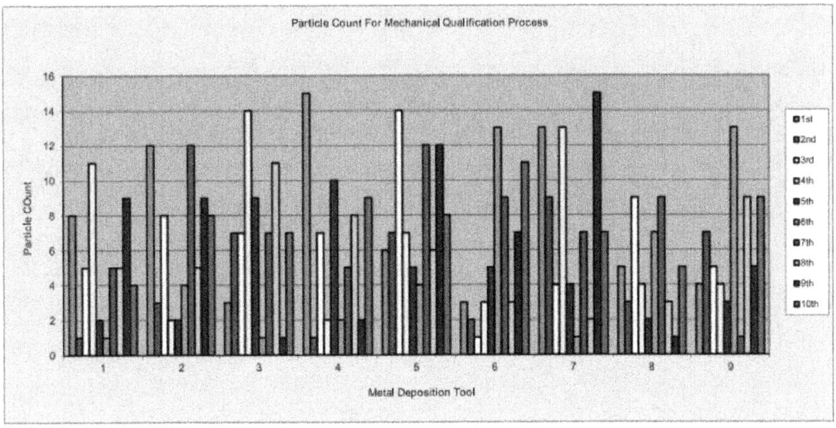

Relationship between shutdown event and particle generation in gas line

Particle sensor is mounted at the gas line as shown in Figure 14 in section 3.7, to monitor particle existence in gas line. Since the gas lines are equipped with in-line gas filters, the particles are trapped in the filter. Zero reading was captured from the particle sensor. From this study, conclusion can be made that particle is not caused by the gas lines.

Relationship between shutdown event and sputter target life

Result for weekly shutdown event versus target life in Kilowatt per Hour (KW/H) is measured (refer Figure 20). By using Minitab, normality test was done for the target life as shown in Figure 21. The result shows the data is normal and valid to be studied.

Correlations test between shutdown and average sputter target life have been done. The Pearson correlation between shutdown and move is 0.981, which means strong correlations between both variables.

Regression analysis was made between shutdown event and sputter target life. Result for R2 value of 96.3% indicates strong relationship between shutdown event and sputter target life. Conclusion can be made that shutdown is highly influenced by sputter target life.

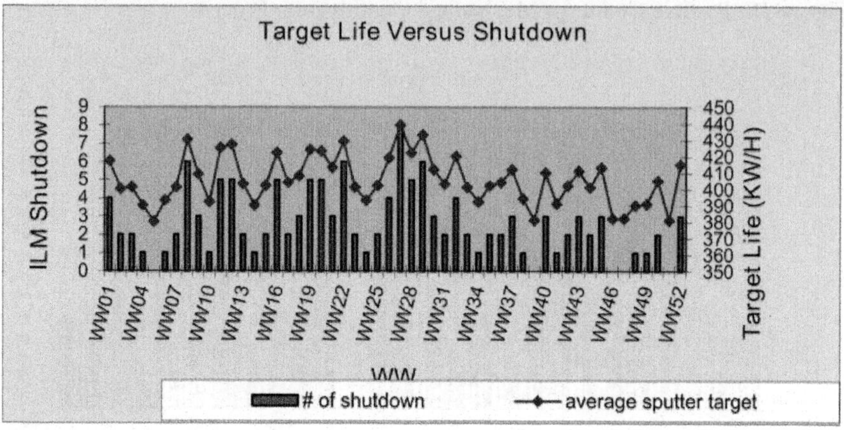

Figure 20: Average Target Life (KW/H) versus ILM Shutdown (weekly)

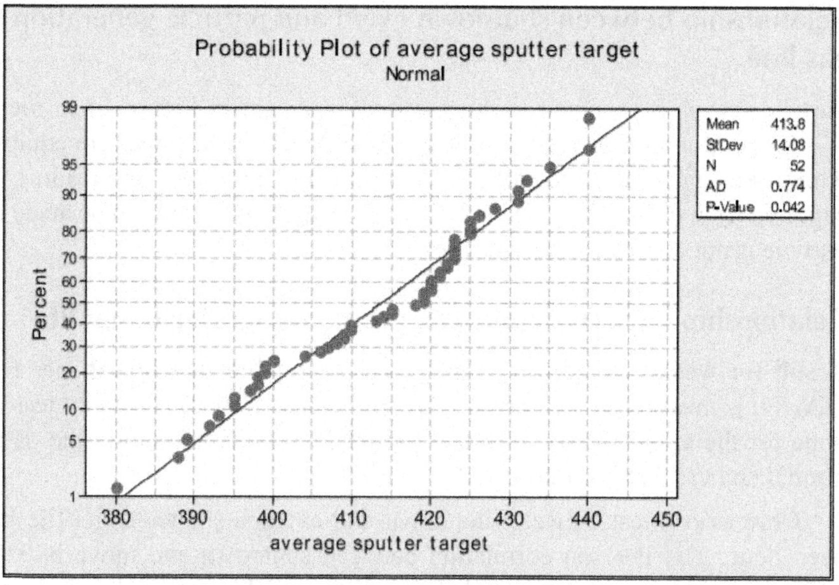

Figure 21: Normality Test for Sputter Target Life

Environment

Result for the load lock environment is shown in Figure 22. Zero lots were captured with particle count more than three in the test. Since the specification is less than three counts, it can be concluded that load lock environment is free from particle.

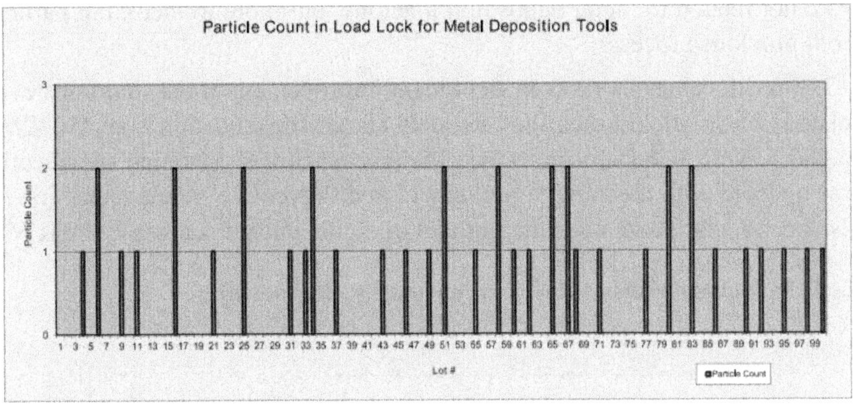

Figure 22: Particle Count in Load Lock at Metal Deposition Machines

ANALYSIS OF TRUE CAUSES

From the verification process of potential true causes, wafer transfer in load lock and sputter target life has the most significant relationship to shutdown event. Since the most significant root cause is the sputter target life, this project will only focus on the improvement of sputter target life.

Sputter Target Life Improvement

From the data of shutdown event versus sputter target life, observation can be made that the rate of machine shutdown is increasing by the increment of sputter target life. Even though the specification for the maximum sputter target life is 450KW/H, however in the study conducted, it shows that the chances of machine shutdown is higher if the sputter target life reach more than 410 KW/H. Zero shutdown per week was observed for sputter target life less than 390 KW/H, ten cases of one shutdown event per week were observed for sputter target life between 390 KW/H- 400 KW/H and 15 cases of two shutdown events per week were observed for sputter target life between 401 KW/H- 410KW/H.

In this study, three machines were selected to be improved by reducing the life of sputter target. Sputter target life is limited to 400 KW/H only, before replace with another new sputter target. Observation of the shutdown event versus the new sputter target life was monitored for three months, and the result is shown in Table 9.

From the data obtained, shutdown event was significantly improved. Three cases of shutdown event were observed within 12 weeks, however the cases were not related to metal deposition machine, but more to incoming particles from previous process.

Since the changes showed significant improvement in the shutdown event related to the particle issues, the new sputter target life reduction from 450 KW/H to 400 KW/H is introduced to the other six machines. Machine specification was updated with this new improvement and Preventive Maintenance job has also been revised to change the sputter target life when it reaches ~ 400 KW/H.

Table 9: Shutdown versus Move base on new Sputter Target Life

WW	Tool			Total	# of shutdown
	1	2	3		
WW01	2,875	2,790	2,698	8,363	0
WW02	2,650	2,923	2,596	8,169	0
WW03	2,748	2,866	2,777	8,391	1
WW04	2,931	3,257	2,956	9,144	0
WW05	3,096	2,385	2,343	7,824	0
WW06	2,386	3,397	2,248	8,031	0
WW07	2,847	2,711	3,464	9,022	1
WW08	3,186	3,290	3,355	9,831	0
WW09	3,727	3,513	3,628	10,868	0
WW10	3,726	3,053	3,972	10,751	0
WW11	3,527	3,331	3,319	10,177	0
WW12	3,459	3,175	3,736	10,370	1

CONCLUSION

Significant improvement can be seen in terms of In-Line Monitoring (ILM) shutdown event after the improvement of new sputter target life. Even though tool shutdown event still appear, it is mainly related to incoming process factors and not due to metal deposition machine. Therefore it is crucial to change the sputter target life when it reaches ~ 400 KW/H.

REFERENCES

1. Alfred, M. 2001. Reduction of Particulate Contamination in Wafer Processing, pp. 220-225. Santa Clara, California: Applied Materials INC.
2. Bates, S.P., 2000. Silicon Wafer Processing, Applied Material Summer, US. http://www.mi.e-atech.edu/jonathan.colton/me4210/waferproc.pdf. Accessed on 24 March 2010.
3. Benjamin, Y.H., Ho-Ahn, L.K. 1987. Particle Technology Laboratory. Mechanical Engineering Department, University of Minnesota, Minneapolis, England.

4. Bordon, P. 1990. The Nature of Particle Generation in Vacuum Process Machines. IEEE Transactions on Semiconductor Manufacturing 3:189-194.
5. Chen, D., Seidel, T., Belinski, S. and Hackwood, S. 1989. Dynamic Particulate Characterization of a Vacuum Load- lock System. J. Vac, Science Technology 7:3105-3111.
6. Chowdhury, N.M. 1997. Designing a Bulk Specialty Gas System for High-Purity Applications. Proceedings of the Institute of Environmental Sciences, pp. 65-72.
7. Cooper, D.W. 1986. Aerosol Science Technology, pp. 25-34. New York: McGraw- Hill. Faraday, M. 1857. The Bakerian lecture: experimental relations of gold (and other metals) to light, Philosophical Transactions of the Royal Society of London 147:145–181.
8. Grove, W.R. 1852. Electro-chemical polarity of gases. Philosophical Transactions of the Royal Society of London 142:87–101.
9. Hamberg, O.1985. Process Annual technical Meeting of Institute of Environment Sciences, Larrabee, G.B.
10. Hart, J. and Paterson, A. 1994. Evaluating the Particle and Outgassing Performance of HighPurity. Electronic-Grade Specialty Gas Cylinders Microcontamination 12:63-67.
11. Hart, J., McDermott, W., Holmer, A. and Natwora J. 1995. Particle Measurement in Specialty Gases, Solid State Technology 38:111-116.
12. Martin, R.W. 1989. Defect Density Measurement, in Proc. 9th International Symposium Contamination, Los Angeles. McDermott, W.T. 1997.
13. A Gas Diluter for Measuring Nanometer-Size Particles in Oxygen or Hydrogen. Proceedings of the Institute of Environmental Sciences and Annual Technical Meeting, pp. 26-33
14. Wu, J.J, Cooper, D.W. and Miller, R.J. 1989. An aerosol model of particle generation during pressure reduction, Journal of Vacuum Science & Technology A: Vacuum, Surfaces, and Films 8:1961-1968.

Chapter 9

A COMPARATIVE STUDY OF THE ECONOMIC FEASIBILITY OF EMPLOYING CHP SYSTEMS IN DIFFERENT INDUSTRIAL MANUFACTURING APPLICATIONS

Chad A. Wheeley, Pedro J. Mago, Rogelio Luck

Department of Mechanical Engineering, Mississippi St at e University, Oktibbeha County, USA

ABSTRACT

Extensive research work including multiple methodologies and numerous simulations have been completed in order to determine the economic effectiveness of employing CHP at commercial and residential sites. In contrast to the above, very few attempts have been made to develop methodologies to study the feasibility of CHP systems at industrial manufacturing facilities. As a result, practical opportunities for CHP at industrial sites are often not realized or even investigated. It follows that there is a need in the CHP related literature for an analysis that is explicit and yet general enough to determine the economic viability and potential for success of CHP systems at industrial manufacturing facilities. Therefore, the purpose of this paper is to clearly outline a methodology to determine the economic effectiveness of installation and operation of a CHP system at industrial facilities that have a need for space or process heating in the form of steam. The effect on the CHP system economic performance of several parameters, such as the project payback, internal rate of return, net present value, etc., are considered in the proposed methodology. The applicability and generality of the methodology is illustrated by examples including four different manufacturing facilities. The effects of the variability of factors such as annual facility operational hours during which both process heat and electricity are needed, facility average hourly thermal load, cost of utility supplied electricity, and CHP fuel type and associated fuel cost, on the outcome of the economic analysis are also examined.

INTRODUCTION

When considering a base-load combined heat and power (CHP) system for an industrial manufacturing facility, a number of different parameters must be examined and addressed before one can determine its estimated economic viability and potential for success. The most widely accepted parameter that is used to estimate the feasibility of any proposed CHP project is known as spark spread, which is essentially the difference in the cost of utility supplied electricity and the fuel cost associated with production of electricity on site [1]. A spark spread of \$12/MMBtu (\$0.041/kWh) is typically considered to be the threshold that is representative of an economically attractive CHP project, meaning that projects that exhibit spark spreads in excess of \$12/MMBtu (\$0.041/kWh) will have a good potential for low payback periods and overall economic success [1]. Graves et al. [2] developed a more sophisticated method that incorporates generator heat rate, thermal recovery efficiency, equipment cost, and acceptable payback period, allowing for a more accurate indication of CHP viability. In a similar manner, Smith et al. [3] developed a detailed model, based on the spark spread, which compares the electrical energy and heat energy produced by a CHP system against equivalent amounts of energy produced by a traditional, or separate heating and power (SHP), system. In addition, they introduced an expression for the spark spread based on the cost of the fuel and some of the CHP system efficiencies as well as an expression for the payback period for a given capital cost and spark spread. However, for industrial manufacturing facilities, in addition to the spark spread, there are other factors that must be considered when analyzing the economic feasibility of a CHP system, such as the type of prime mover, the fuel availability and cost, operation hours, among others. Typically prime movers used in manufacturing facilities include, but are not limited to: steam turbines, combustion turbines and internal combustion engines. Reciprocating engine and fuel cell CHP systems are other options that could possibly be considered for industrial manufacturing facilities. However, these technologies are often expensive and have somewhat limited operating ranges. Micro-turbines are a good choice for smaller commercial and residential buildings, but they typically do not have the capacity to offset an adequate amount of an industrial manufacturing facility's base electrical load. Ellis and Gunes [4] presented a comparison of different generating system characteristics, which addressed the use of fuel cells. Steam turbines are frequently employed due to their fuel flexibility as well as their ability to provide an extensively wide range of process steam supply flow rates when compared to combustion turbines. For example, combustion turbine CHP units are typically rated to supply a certain amount of steam, with one or two increased steam flow rate options available if duct burners are added. However, steam turbines, on the other hand, allow for multiple variations in process steam flow rates [5]. Thus, the desired process steam flow rate can be attained by a number of different methods, such as utilization of extraction steam turbines instead of backpressure steam turbines or by optimization of the backpressure turbine boiler system, which can be easily modeled by making use of the US Department of Energy's Steam System Assessment Tool (SSAT) [6] or any other appropriate turbine modeling software.

Combustion turbines, on the other hand, are often more easily integrated into an industrial facility's operating scheme. Also, as will be seen in one of the cases presented in the comparative analysis section of this paper, a combustion turbine CHP system can often allow for positive electrical cost savings, which is seldom the case for steam turbine CHP systems. In addition, the use of renewable fuels is on the rise due to the price surge and volatility of traditional fuels, as well as a general desire to decrease on site emissions and use more environmentally friendly fuel sources. For example, biomass, such as waste materials from agricultural or industrial processes that are available at or close to the CHP site and sometimes free of charge can be a cost effective CHP fuel source which can be used to generate heat and power for a manufacturing facility [7].

Modeling of CHP system has been extensively investigated for commercial buildings [8-16]. However, very little research has been performed on CHP for the industrial sector and very few and methodologies have been developed to evaluate the performance of these types of systems at industrial manufacturing facilities [17]. Therefore, this paper presents a detailed model which can be used to evaluate the economic performance of a CHP system at an industrial manufacturing facility. The model presented in this paper calculates the cost savings, if any, associated with the particular system used, payback period, internal rate of return, and net present value, of the proposed CHP project. The proposed model can be applied to any manufacturing facility and allows for analysis of different CHP prime movers and system configurations. In order to illustrate how the proposed model can be applied to any manufacturing facility, four different industrial sites were selected as case studies.

In general, there are a number of parameters that play a vital role in the outcome of the economic analysis of a CHP system. Therefore, these factors can often be used to gauge the economic attractiveness of any such CHP system. However, since each of these parameters can vary greatly from one facility to the next, the model developed in this paper was applied to multiple cases in order to illustrate not only how each of these factors can provide insight to economic considerations of any such CHP system but also how the model assesses variations in these parameters. The factors which are analyzed in this paper are the annual operating hours of the facility during which both electricity and process heat are required (equivalent to the annual operating hours of the CHP system), the usage rate of conventionally supplied electricity, the average hourly thermal load of the facility, and finally the CHP system fuel type and its associated fuel cost.

ANALYSIS

The following section presents a methodology that can be used to conduct an economic analysis and feasibility study for a CHP system to be installed at an industrial manufacturing facility. It is important to note that the methodology developed in this section is only to be applied for CHP systems considered at industrial facilities that have a need for space or process heating in the form of steam. If thermal energy is to be supplied in another form, the methodology must be modified. Step 1: Estimate the size of the CHP power generation unit (PGU) using information from the monthly utility bills and/or information regarding the steam requirements of the facility. It is recommended to initially size the system based on the minimum monthly demand and then modify the PGU size to obtain the best economic performance. Another option is to select a PGU to supply all the steam requirements of the facility. However, sizing the PGU to supply the facility's entire steam load can result in the production of excess electricity. This outcome is not preferable for regions that have an unfavorable net metering incentive, which is the case for many of the southeastern United States. Therefore, the capacity of the system can be expressed as:

$$Cap_{sys} = L_e \tag{1}$$

or

$$Cap_{sys} = (PGU)_{steam\text{-}req} \tag{2}$$

where L_e initially is the PGU size based on the minimum monthly demand, which is then modified to obtain the best economic performance, and $(PGU)_{steam\ req}$ - is the PGU size obtained after supplying the optimum amount of the facility's steam requirement.

Step 2: Determine the installation cost. In this step, it is important to note that some of the equipment needed to convert to CHP may already be in place and thus will only need to be retrofitted. The installation cost (IC) can be determined as:

$$IC = (CR)(Cap_{sys}) \tag{3}$$

where CR is the cost rating which can be obtained from the EPA Catalog of CHP Technologies [18].

Step 3: Determine the system's annual electrical production as follows:

$$Prod = \left(Cap_{sys}\right)\left(Hr\right)\left(AF\right) \tag{4}$$

where Hr is the CHP unit's annual operating hours, which is equivalent to the operating hours of the facility during which electricity and process heat are both required, and AF represents the estimated availability factor of the CHP unit. The AF is included in order to account for the fact that the proposed system will most likely experience periodic downtime either due to trips or for scheduled maintenance. This value can be easily varied and modified as desired.

Step 4: Determine the operation and maintenance cost associated with running the CHP unit. The EPA CHP Catalog recommends an operation and maintenance (O&M) rating of \$0.005/kWh for steam turbine CHP units. However in order to account for any maintenance fees that result from the additional CHP equipment (i.e. boiler, ductwork, etc.) an O&M rating of \$0.008/kWh will be used for the steam turbine cases considered in this analysis. Therefore, the annual O&M cost can be obtained using the annual production and the operational and maintenance cost rating, $O\&M_{rating}$:

$$O\&M = \left(Prod\right)\left(O\&M_{rating}\right) \tag{5}$$

Step 5: Estimate the annual CHP system operational ($Cost_{op}$), resulting from the CHP unit's fuel consumption. The CHP system operational cost can be evaluated as:

$$Cost_{op} = \left(fuel_{FR}\right)\left(cost_{f}\right)\left(Hr\right)\left(AF\right) + \left(O\&M\right) + lost_{rev} \tag{6}$$

where $fuel_{FR}$ is the CHP unit fuel feed rate, $Cost_f$ is the fuel cost, and $lost_{rev}$ is any revenue that might be lost due to operation of the CHP system. For instance, facilities that produce waste streams which can be utilized as a fuel source often sell this waste to fuel suppliers. If this waste stream is considered as the CHP system fuel, it can no longer be sold for profit and the loss in revenue due to this action must be accounted for. rev lost The loss in revenue can be calculated as

$$lost_{rev} = \left(fuel_{cons}\right)\left(SR\right) \tag{7}$$

where $fuel_{cons}$ is the annual CHP unit waste fuel consumption and SR is the sale rate, which is the rate at which waste was sold by the facility. If there is no loss in revenue, then $lost_{rev}$ should be set to \$0.00. The FR fuel can be obtained in different ways: 1) from the manufacturer; 2) using the information of the PGU efficiency, or c. using SSAT software [6] to model the selected PGU.

Step 6: Determine the usage rate of electricity produced by the CHP unit (CHP_{UR}) as follows:

$$UR_{CHP} = Cost_{op} / Prod \tag{8}$$

Step 7: Estimate the potential electrical cost savings resulting from operating the CHP system, which is based on the difference between and the cost of utility supplied electricity . (CS_{ele}) UR_{CHP} (UR_{conv})

$$CS_{ele} = (Prod)(UR_{conv} - UR_{CHP}) \tag{9}$$

In Equation (9), a negative CS_{ele} value implies that the CHP system does not provide any cost savings based on electricity alone.

Step 8: Estimate the thermal energy cost savings associated with offsetting a portion or the facility's entire process heating load. First it is necessary to determine the thermal energy savings resulting from operation of the CHP system (ES_{st})

$$ES_{st} = (Ld_{st})\left(\frac{29.9\ boiler\ hp}{1000\frac{\text{lb}}{\text{hr}}\ steam}\right)\left(\frac{33,479\frac{\text{Btu}}{\text{hr}}}{boiler\ hp}\right) \times \left(\frac{\text{MMBtu}}{10^6\ \text{Btu}}\right)(Hr)(AF) \tag{10}$$

where Ld_{st} is the portion of the facility's process heating load (portion of the steam flow rate) that is to be offset by steam produced from waste heat recovered by the CHP system and the other values used in the above equation are typical conversion constants. The thermal energy (steam) cost savings (CS_{st}) is then the product of the thermal energy savings and the usage rate of conventionally supplied thermal energy (UR_{th}), taking into account any associated boiler efficiency (η_{boiler}) values.

$$CS_{st} = (ES_{st} / \eta_{boiler})(UR_{th}) \tag{11}$$

Step 9: Estimate the total annual project cost savings (CS_{tot}) as

$$CS_{tot} = CS_{ele} + CS_{st} + Rev_{gen} \tag{12}$$

where Rev_{gen} accounts for any additional revenue that might be generated by the sale of a waste fuel source that is now unused due to operation of the CHP system. For instance, if an industrial facility is utilizing a waste stream as a fuel source for process heat and the proposed CHP system offsets a portion of this waste fuel, the portion which is now unused could be sold to fuel suppliers or other facilities that utilize that particular type of fuel. Any additional revenue generated by the sale of a waste fuel source that is now unused due to operation of the CHP system can be calculated as:

$$Rev_{gen} = (fuel_{avail})(SR) \tag{13}$$

where $fuel_{avail}$ is the fuel that could be sold as a result of operating the CHP system. If there is no revenue generated by the sale of a waste fuel then Rev_{gen} should be set to \$0.00

The project simple payback (SP) is then calculated as

$$SP = IC/CS_{tot} \tag{14}$$

The internal rate of return (IRR) is obtained by applying a numeric solver to the following implicit equation

$$-IC + \sum_{n=1}^{lc-year} CS_{tot}/(1+IRR)^n = 0 \tag{15}$$

where lc-year represents the number of year life cycle. The solution to the above equation, i.e., IRR, is usually available in spreadsheet software applications

The net present value (NPV) is calculated as

$$NPV = (-IC)(1-ITC\%) + \sum_{n=1}^{lc-year} CS_{tot}/(1+ir)^n \tag{16}$$

where ITC% is the percentage of the implementation cost that is covered by the Investment Tax Credit and ir is the interest rate that the facility in question could receive had it invested the capital in another venture rather than using it to fund the CHP project.

Step 10: After obtaining an initial economic performance, different PGU types and sizes can be evaluated to determine the optimum size and technology that provides the best economic outcome.

Table 1: Energy load and operational data for the selected facilities

facility	*base electric load* (kw)	*thermal load* (mmbtu/hr)	*power to heat ratio*	*annual operating hours** (hr/yr)
Case 1: Food Products Rendering Plant	4600	213.8	0.074	6864
Case 2: Lumber Mill	3200	27.3	0.401	2750
Case 3: Plastics Manufacturing Plant	15,000	29.8	1.717	7008
Case 4: Chemical Plant	10,000	18.5	1.842	8760

RESULTS AND DISCUSSIONS

In order to illustrate how the methodology presented in Section 2 may be used to determine the economic viability of installing a CHP system at a particular industrial manufacturing facility, a number of economic analyses for CHP units at different manufacturing plants are considered. The proposed industrial facility CHP projects considered in this section were chosen to illustrate a wide range in facility operational inputs used in each economic analysis and all of the facilities considered have a need for both electricity, which is currently provided by local utilities, and thermal energy in the form of process steam. Each of the facilities considered manufacture different products, have significantly different electrical and thermal loads, have different annual operating hours, and some even have available on-site fuel sources. The facilities considered in each case were chosen based on these variations in order to add robustness to the analysis as well as to illustrate how the methodology can be applied to a number of different industrial facilities which differ from one another. Table 1 presents the base electrical and thermal loads (before considering CHP), the power to heat ratio (ratio of the electric to the thermal load), and also the annual operating hours for each of the selected facilities.

In all the calculations a 10-year life cycle and 15% interest rate that the facility in question could receive had it invested the capital in another venture rather than using it to fund the CHP project were considered. In addition, an estimated AF of 0.8 is used for all of the analysis included in this paper. While this value may seem high, it helps to ensure that any conclusions made remain conservative.

Economic Performance of the Evaluated Cases

The first three cases presented in this section were analyzed using a steam turbine, while the last case was analyzed using a combustion turbine in an effort to establish the differences between these two types prime movers. Case 1: The first case presented analyzes a backpressure steam turbine CHP system proposed for a food products rendering plant located in central Mississippi. The facility considered in Case 1 operates for 6864 productions hours per year during which both electricity and process heat are required. The most economical CHP option considered for the facility was a backpressure steam turbine CHP unit fueled by biomass. The PGU was selected to supply all the steam required by the facility (156,200 lb/h), which resulted in a 3.46 MW electricity capacity.

Case 2: This case analyzes a backpressure steam turbine CHP system proposed for a lumber facility located in northern Mississippi. The facility considered in this case operated for 2750 production hours per year during which both electricity and process heat were required. The most economical CHP option considered for this facility was a backpressure steam turbine CHP unit, which was sized using the SSAT software [6] and the facility's average base electric load (3200 MW). However, for this case, the facility generated a large amount of wood waste on-site and sold it to local biomass suppliers in order to generate additional revenue. The most economical CHP system for the facility required that a large portion of this wood waste no longer be sold but rather be utilized as fuel for the CHP unit. Therefore, there is lost revenue associated with this case. The facility considered in this case also used a large portion of another waste stream, planar wood shavings, as a fuel source for wood fired boilers which supplied process heat in the form of steam to the wood drying kilns. The CHP system considered provided the facility with the capability to offset a portion of this steam.

As a result, a portion of the wood fuel that was supplied to the existing boilers was no longer used and could then be sold to the same local biomass fuel suppliers, resulting in an additional generated revenue source. Case 3: Case 3 analyzes an extraction steam turbine CHP system that was proposed for a plastic products manufacturing facility located on the Mississippi Gulf Coast. For this case, a natural gas fueled boiler/steam turbine CHP unit was considered which were also sized using the SSAT software [6] and the facility's base electric load. The facility analyzed in this case operates for 7008 hours during the year. Case 4: As mentioned before, to establish a contrast between steam turbines and combustion turbines in CHP applications, another case that utilizes a combustion turbine is included in this paper. Case 4 presented a CHP system proposed for a chemical manufacturing facility on the Mississippi Gulf Coast. The most economical option considered for this facility was a 5.7 MW combustion turbine CHP system. The facility's annual base electric load was used to determine which combustion turbines would supply an adequate amount of electricity as well as process heat. Based on the facility's needs, three different sizes of combustion turbines were considered and analyzed using equipment specifications provided by the combustion turbine manufacturer and the most economically viable option was chosen.

The facility considered in Case 4 operates for 8760 production hours annually. The O&M cost for this case was zero since a combustion turbine CHP unit was utilized and the equipment manufacturer provided a system warranty which covered maintenance fees. The methodology was applied to each of the four cases described in Table 1 and the results obtained in each step are presented in Table 2. From Table 2, it can be observed that Case 1 exhibits a favorable CHP system economic performance. The facility considered in Case 1 has a very large process heating load and a low PHR (0.074). In addition, it also has a relatively large amount of annual operating hours (≈78% of the time during a year), which allowed for longer CHP system operation. The annual electrical consumption which was to be offset by the CHP system considered for this case was somewhat large and the associated CHP electrical production rate was relatively high. Therefore, the cost of producing only electricity from the CHP system was more expensive than purchasing conventional electricity from the grid. However, the thermal load which was to be offset by the CHP system for this case was relatively high, resulting in high thermal energy cost savings. This was able to adequately counter the increase of the electrical cost from operation of the CHP unit, which resulted in an economically attractive project.

Therefore, this case illustrates how a low PHR combined with a large amount of annual operating hours yields good annual cost savings and therefore a good payback period. Case 3 on the other hand had a somewhat large electrical base load but a relatively small process heating load, which yielded a high PHR (1.717). Table 2 illustrates that even though the annual facility operational hours during which the CHP system was to be utilized were high for this case (≈80% of the time during a year), there were no cost savings and therefore the use of a CHP system was not economically feasible. This was mostly due to the combination of the high electrical usage and low thermal usage which were to be offset by the CHP unit. As a result, the low thermal energy cost savings were incapable of countering the increase in electrical cost from CHP.

Table 2: Methodology results for the four evaluated cases.

Methodology	*Case 1*	*Case 2*	*Case 3*	*Case 4*
		Step 1		
Cap_{sys} [MW]	3.463	0.63	15.45	5.7
CR [\$/kW]	2900	2900	1100	1313
		Step 2		
IC [\$]	10,042,700	2,661,820	16,997,200	7,484,100
HR (hours)	6864	2750	7008	8760
AF	0.8	0.8	0.8	0.8
		Step 3		
Prod [MWh/yr]	19,016	1386	86,630	39,945
		Step 4		
O&M [\$/yr]	152,128	11,088	693,040	0
$Lost_{rev}$ [\$/yr]	0	118,800	0	0
$cost_f$	\$21.00/ton	\$0.00/ton	\$4.510/MMBtu	\$4.421/MMBtu
$fuel_{FR}$	25.8 tons/hr	4.5 tons/hr	312.7 MMBtu/hr	61.0 MMBtu/hr

		Step 5		
$Cost_{op}$ [$/yr]	3,127,260	129,888	8,599,617	1,889,924
		Step 6		
UR_{CHP} [$/kWh]	0.16445	0.09371	0.09927	0.047312
UR_{conv} [$/kWh]	0.0825888	0.05497	0.0732886	0.061793
		Step 7		
CS_{ele} [$/yr]	−1,556,674	−53,693	−2,250,771	578,434
Ld_{st} [lb/hr]	156,200	27,222	22,000	18,500
		Step 8		
ES_{st} [MMBtu/yr]	858,602	59,949	123,467	129,780
CS_{st} [$/yr]	4,007,096	106,531	556,835	675,011
Rev_{gen} [$/yr]	0	97,092	0	0
		Step 9		
CS_{tot} [$/yr]	2,450,421	149,929	-1,693,935	1,253,445
lc-year [yr]	10	10	10	10
ITC%	10%	10%	10%	10%
SP [yr]	3.69	15.98	N/A	5.37
IRR	23.94%	N/A	N/A	13.24%
NPV [$]	3,259,668	−1,643,176	N/A	−444,937

Case 3 is a good example that a high PHR is a parameter that may indicate that a CHP system may not be economically feasible for that particular facility despite the fact that the CHP system could be utilized for a high amount of annual operating hours and the system installed cost rating ($/kW) was the lowest for all of the cases considered.

Case 2 differed from all of the other cases considered in that the fuel needed to operate the proposed CHP system was generated on site as a waste stream. However, this waste fuel was sold by the facility to local biomass fuel suppliers, so any amount that was to be utilized as a CHP system fuel source resulted in a loss in revenue for the facility. The thermal load for this case was also relatively small, which yielded a low PHR (0.041). However, the thermal energy cost savings was still adequate to counter the associated electrical cost increase from use of the CHP system considered in this case. On the other hand, the annual facility operating hours during which both process heat and electricity were needed were very low. Therefore, the proposed CHP unit only operated 2750 hours annually (31% of the time), which significantly decreased its capability to provide overall project cost savings. The low operating hours of the proposed CHP unit along with the associated revenue loss related to utilization of the waste fuel ultimately resulted in a poor economic performance and a relatively long project payback period for this case.

In general, for the cases that employed steam turbines (1, 2, and 3), the electricity production from the CHP system was more expensive than the electricity produced using conventional technologies. However, if the thermal load which was to be offset by CHP system is relatively high, the thermal energy cost savings can counter the increase of the electrical cost from the CHP operation, resulting in an economically attractive implementation. On the other hand, if the thermal load to be offset by the CHP unit is small, the thermal energy cost savings will be low and will most likely result in poor overall project cost savings. This can be clearly seen in Equation (11), in which the cost savings associated with the thermal load and any revenue that might be generated by the sale of a waste fuel source that is unutilized due to operation of the CHP system attempt to balance the negative cost savings typically associated with generation of electricity on site.

Comparison of steam turbine prime movers to combustion turbine prime movers for industrial facility CHP systems.

Case 4 analyzed a CHP system for a chemical manufacturing plant that had an average base electrical load but a relatively small process heating load, which in turn yielded a high PHR (1.842). However, rather than analyzing a steam turbine, a combustion turbine CHP system was considered for this case. The facility considered in this case operated for 8760 hours per year (non-stop) and the resulting CHP electrical production rate was lower than the conventional electrical purchase rate, meaning that there were electrical cost savings resulting from use of the CHP unit, which is seldom the case for a steam turbine CHP system. The resulting annual electrical cost savings was still somewhat low. The corresponding thermal energy cost savings was also relatively low due to the facility's low process heating load which was to be offset by the CHP system. However, much of the equipment needed for the CHP project was already installed or could easily be retrofitted and much of this equipment was not being utilized to its full potential. As a result, the CHP system installation cost was very low. Therefore the use of a CHP system for this case exhibited good economic considerations in spite of the fact that the annual cost savings were lower for this case than for many of the other cases considered.

It is important to highlight that Case 4 is the only case in which the cost of the electricity produced by the CHP system is lower than the conventional electrical cost. However, when using a combustion turbine, it is important to note that the ability to significantly vary the CHP system steam supply rate will be greatly decreased. For instance, the steam supply rate for a steam turbine CHP system can be relatively easily increased or decreased over a wide range by modifying the boiler fuel input and boiler steam flow rate. Typically, combustion turbine CHP systems are rated to recover a certain amount of heat from the exhaust and utilize that heat source for process steam production. If additional steam is required by the facility, then the combustion turbine CHP system can often be equipped with a duct burner, which requires additional fuel input in order to produce excess steam. However, duct burners that are incorporated into combustion turbine CHP systems are usually only available in two or three sizes, thus limiting the options for increasing the process steam flow rate. The reduced capability to modify the CHP process steam flow rate is an important aspect that must be thoroughly addressed when considering a combustion turbine CHP application. For instance, it is often the case that a facility could generate electricity at a rate lower than the conventional utility electrical cost if they utilize a combustion turbine as the prime mover for a CHP system they are considering. However, the thermal energy cost savings might be substantially less than the thermal energy cost savings associated with a steam turbine CHP system due to the steam supply flow rate restrictions corresponding to the combustion turbine. Therefore, combustion turbines may not always be the most economically attractive option. For instance, in many cases, the increased thermal energy cost savings resulting from utilizing a steam turbine CHP application could outweigh the electrical cost savings benefits of a combustion turbine. Another aspect that influences the economic performance of a CHP system is the annual operating hours.

In general, it is apparent that longer system operational hours result in better economics for the use of CHP systems. From the results presented in Table 2, it can be concluded that some of the key parameters to be considered during a CHP project economic analysis are the PHR (electric and thermal loads), the annual operating hours, the electric utility rates, and of course the cost and availability of the fuel to be used to operate the CHP system. For this reason the following section evaluates how varying some of these parameters will affect the economic performance of CHP systems.

Parametric Analysis of Some of the Factors

That Affect the CHP System Performance This section presents the effect of several parameters on the economic performance of CHP system for the evaluated cases. These parameters include: annual facility operating hours, electric utility usage rates, the facility electrical and thermal load (represented by the PHR), and the fuel to be used to operate the CHP system.

Annual Facility Operating Hours

CHP systems are often good alternatives for industrial manufacturing facilities that require both electrical power and process heat. However, these projects will not result in good economics if the CHP units are operated during times when only electricity or only process heat are required by the facility in question.

Therefore, the annual facility operating hours during which both electricity and process heating are required is an important parameter that has a significant impact on the economic success of a CHP project. To assess the effect of the operating hours on CHP economic performance, the facilities were evaluated using 8760 hr, 6570 hr, 4380 hr, and 2190 hr, while all of the other independent parameters, such as their corresponding base electric loads, thermal loads, etc. are held constant. Figure 1 shows the effect of the operational hours on the CHP system economic performance for all the evaluated cases. Figure 1(a) illustrates that for Cases 1, 2, and 4 increasing the hours of operation increases the annual cost savings obtained from the CHP system. This is due to the fact that larger portions of the facilities electrical and thermal energy usages are offset by their respective CHP systems as the CHP operating hours are increased. While this does mean that in some cases the CHP electrical energy cost will be higher, the associated thermal energy cost savings will also be higher, which provides a better potential for improved overall project economics. However, for Case 3, increasing the CHP operational hours represents a decrease in the already poor economic performance.

For this case, the electrical cost resulting from operation of the CHP system is higher than the conventional system electrical cost. Also, this facility (Case 3) requires a relatively low steam flow rate to offset all of the process heating requirements. The annual thermal energy cost savings are far too low to offset the negative electrical savings when the normal facility operating hours (7008 hr/yr) are used in the economic analysis and even when the facility operating hours are increased to a maximum (8760 hr/yr), the total CHP system project cost savings remain negative for Case 3. Figure 1(b) illustrates the simple payback for different operating hours for the evaluated facilities. The results presented in this figure agree with the results obtained previously that are presented in Figure 1(a) since it is the case that greater annual savings yield lower payback periods. The payback time period was not applicable for Case 3 since the CHP system considered for the facility in question exhibited no cost savings.

Facility Electric Utility Rate

Another important parameter that strongly affects the economic performance of a CHP system is the facility's local electric utility rate for purchase of conventionally supplied electricity. To assess the effect of the facility electric utility rate on the CHP systems' economic performance, the facilities considered in Cases 1-4 were evaluated using assumed electric utility rates of \$0.050/ kWh, \$0.075/kWh, \$0.100/kWh, and \$0.125/kWh, while all of the other independent parameters such as the base electric load, thermal load, operating hours, etc.

are held constant. Figure 2(a) illustrates the concept that higher electric utility rates result in higher annual cost savings that are associated with operation of a CHP system. Favorable economics are obtained for Case 3 as the electric utility rate is increased above $0.095/kWh. Figure 2(b) shows that the payback for cases 1, 2 and 4 decreases as the electric utility rate is increased, which is the expected result. However, for Case 3, payback values only become applicable after the $0.095 electric utility rate threshold is exceeded. Even though there are some cost savings associated with the CHP system considered for Case 3 after the $0.095 electric utility rate threshold was exceeded, the corresponding payback is still extremely high. This is why it is significantly important to analyze both the cost savings and the payback period for the implementation of a CHP system. Therefore, it is apparent that the electric utility rate has a strong influence on the economic feasibility of a CHP system.

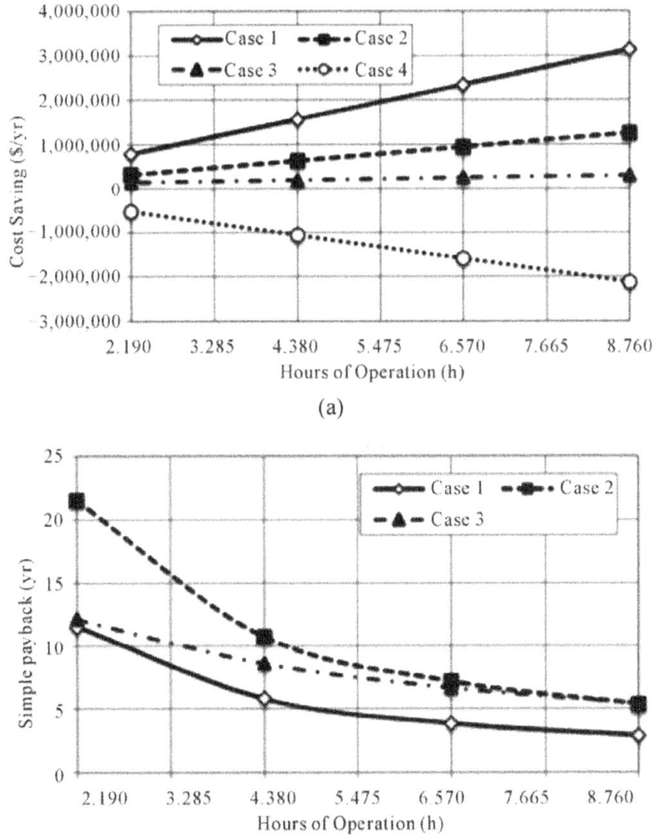

(a)

(b)

Figure 1: Effect of the annual operating hours on (a) the annual cost savings (b) the simple payback

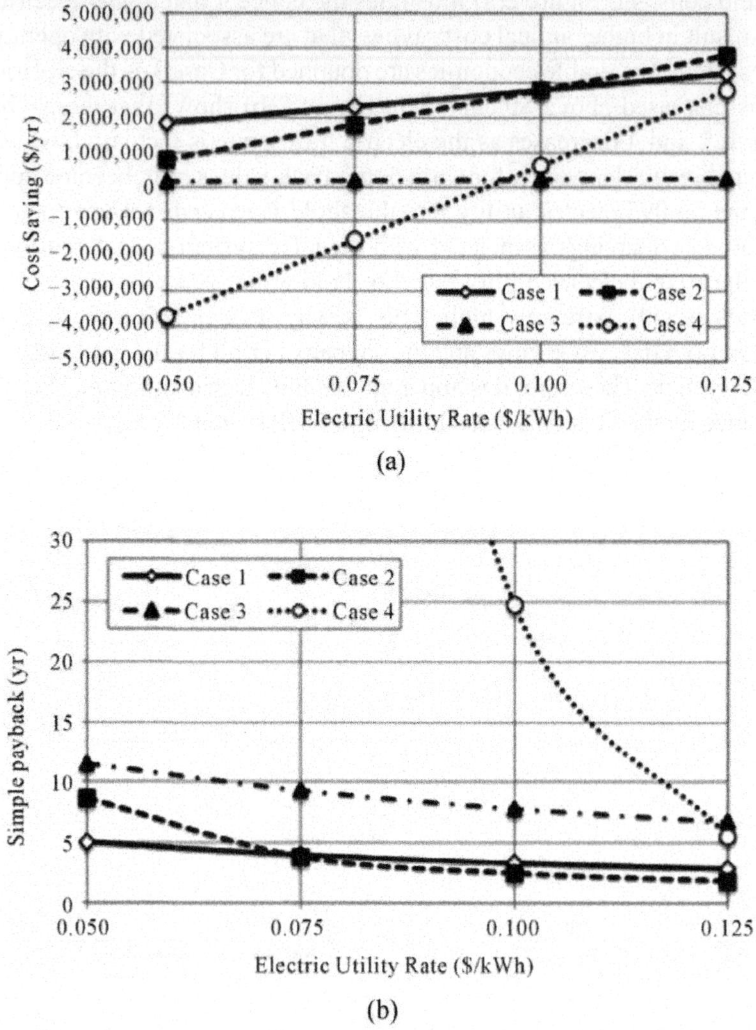

Figure 2: Effect of the electric utility rates on (a) the annual cost savings (b) the simple payback.

Facility Thermal Load

The thermal load of facilities for which CHP systems are proposed is another important parameter that has a significant impact on the economic success of a CHP project. This can also be evaluated as the effect of the power-to-heat ratio (PHR) on the economic performance of the CHP system. The PHR can be expressed as the ratio of the facility's base electric load to its hourly thermal load. To evaluate the effect of the facility's thermal load on the economic performance of a CHP system, the thermal loads of each of the facilities considered in Cases 1-4 were decrease by 25% and 50% and also increased by 25%, while all of the other independent parameters, such as the base electric load, operating hours, etc., were held constant. Figure 3 shows the effect that varying the thermal load has on the annual cost savings and the payback period. Figure 3(a) illustrates that for cases 1, 2, and 4, higher the thermal loads, or in other terms smaller PHRs, will result in greater cost savings associated with operation of the CHP systems. However, the thermal load would have to be unrealistically increased to obtain cost savings for Case 3 due to the extremely poor total cost savings for this case. This can be realized by examining the trend for Case 3 in Figure 3(a). As the thermal load is varied from 50% to 125%, there are minimal changes in the cost savings associated with the CHP project considered for Case 3 and it is also apparent that the thermal load would have to be increased greatly before positive project cost savings would be obtained.

Fuel Selection and Cost

The fuel selection, cost, and availability of the fuel to be used to operate the CHP system are very important factors to consider when determining the economic performance of a CHP system. Figure 4 shows the annual cost savings as well as the payback period for different CHP fuels used for the facility evaluated in Case 1. The fuels used in this case are: typical green wood, natural gas, number 2 fuel oil, and typical western coal. In addition, the costs of the evaluated fuels, which are obtained from the SSAT software [6] estimates, are presented in Figure 4. The fuel energy required in the boiler to satisfy the steam requirements of the evaluated facility is about 271 MMBtu/h.

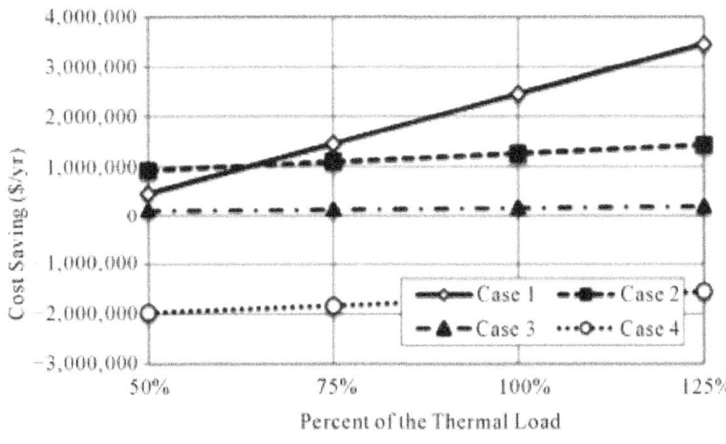

(a)

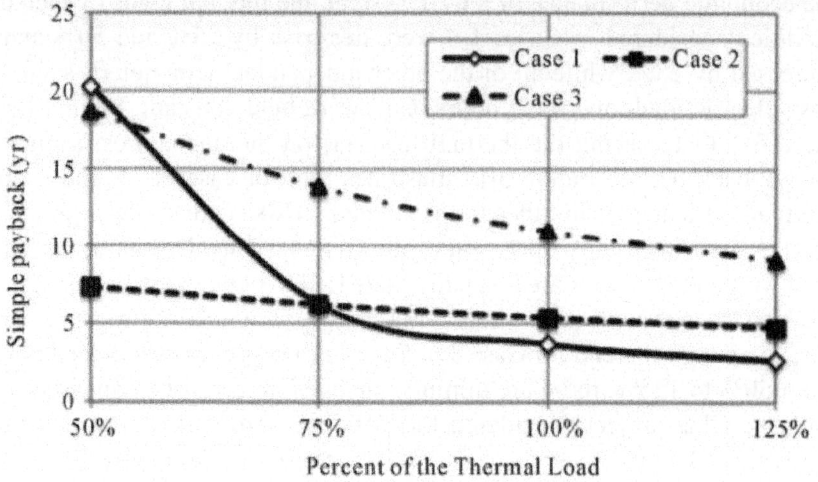

(b)

Figure 3: Effect of the facility thermal load on (a) the annual cost savings (b) the simple payback.

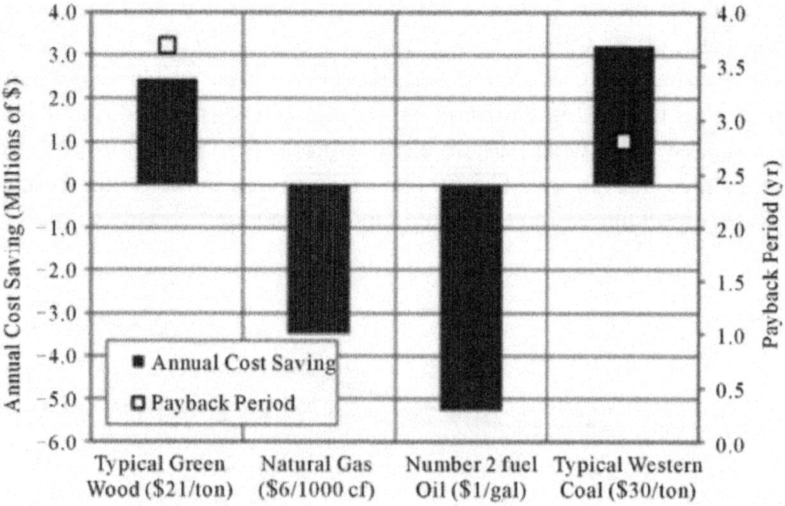

Figure 4: Effect of the CHP fuel used on the cost savings and payback period for the facility analyzed in Case 1.

Therefore, the amount of fuel needed will depend on the specific fuel's heating value. Figure 4 illustrates that using typical green wood and typical western coal provide annual cost savings and paybacks on the order of $2.4 M and 3.69 yr and \$3.2 M and 2.81 yr, respectively. On the other hand, natural gas and number 2 fuel oil both provide negative cost savings, or annual costs which exceed their respective conventional costs. The results presented in this figure show how important the fuel selection is in relation to the economic performance of a CHP system. However, it is also important to keep in mind that the fuel selection is often driven by the availability of the particular type of fuel at the desired location and that the region where the facility is located will impact the cost of the fuel as well.

CONCLUSIONS

This paper presented a methodology which can be used to conduct a feasibility study and economic analysis for a CHP system at an industrial manufacturing facility that has a need for space or process heating in the form of steam. While numerous methodologies have been developed and countless simulations have been completed for CHP systems at commercial and residential buildings, the methodology developed in this paper is highly valuable as it allows for identification of favorable CHP projects at manufacturing plants.

The methodology allowed for analysis of multiple parameters that are indicative of favorable economic performance for CHP and also accounted for any variations encountered due to differing availability of resources, energy requirements, or operating schemes of the facility considered. The effects that variations in many of these indicative factors, such as annual facility operational hours during which both process heat and electricity were needed, facility average hourly thermal load, the cost of utility supplied electricity, and the CHP fuel type and associated fuel cost, have on the outcome of the economic analysis were also examined. Four cases studies were analyzed in order to determine how each of the factors mentioned previously affect the economic considerations of installing a CHP system. In general it was observed that CHP systems that had high annual operational hours resulted in favorable economics and facilities that required less process heat exhibited poor economics when compared to the other cases. Also, it was observed that CHP economics could possibly be improved if a facility was able to utilize a waste stream produced on site as a fuel source for the CHP system. However, variations in the other parameters can negatively counter any of these available benefits and therefore all of the indicating factors must be thoroughly analyzed when conducting a CHP feasibility study. In general, the project payback timeline was decreased and both the internal rate of return and net present value were increased as 1) the operational hours during which both process heat and electricity were required by the facility were increased; 2) the average hourly thermal load of the facility was increased; and 3) the cost of utility supplied electricity was increased. The type of fuel to be used in the CHP unit had a significant impact on the economic performance of the system.

From the case considered, it was observed that some of the evaluated fuels provided favorable economic analysis results while other fuels resulted in negative annual cost savings. Therefore, in order to add robustness to any CHP feasibility study, it is apparent that multiple fuel types should be considered when determining the system's economic performance

REFERENCES

1. J. J. Cuttica and C.Haefke, "Combined Heat and Power (CHP): Is It Right for Your Facility?" 2009. http://www1.eere.energy.gov/industry/pdfs/webcast_2009-0514_chp_in_facilities.pdf
2. R. Graves, B. K. Hodge and L. M. Chamra, "The Spark Spread as a Measure of Economic Viability for a Com- bined Heating and Power Application with Ideal Loading Conditions," Proceedings of the ASME 2008 2nd Con- ference on Energy Sustainability, Jacksonville, 10-14 August 2008, Paper No. ES2008-54203, pp. 167-171.
3. A. Smith, N. Fumo and P. Mago, "Spark Spread—A Scree- ning Parameter for Combined Heating and Power Sys- tems," Applied Energy, Vol. 88, No. 2, 2011, pp. 1494- 1499. doi:10.1016/j.apenergy.2010.11.004
4. M. Ellis and B.Gunes, "Status of Fuel Cell Systems for Combined Heat and Power Applications in Buildings," ASHRA Transactions, Vol. 108, 2002, pp. 108-111.
5. G. Zimmer, "Modeling and Simulation of Steam Turbine Processes: Individual Models for Individual Tasks," Ma- thematical and Computer Modeling of Dynamical Systems, Vol. 14, No. 6, 2008, pp. 469-493. doi:10.1080/13873950802384001
6. U.S. Department of Energy, "Steam System Assessment Tool (SSAT)," 2010. http://www1.eere.energy.gov/industry/bestpractices/software.html
7. Resource Dynamics Corporation, "Combined Heat and Power Market Potential for Opportunity Fuels," 2004. http://www1.eere.energy.gov/library/
8. R. Zogg, K. Roth and J. Brodrick, "Using CHP Systems in Commercial Buildings," ASHRAE Journal, Vol. 47, No. 9, 2005, pp. 33-36.
9. F. A. Al-Sulaiman, F. Hamdullahpur and I. Dincer, "Trigeneration: A Comprehensive Review Based on Prime Movers," International Journal of Energy Research, Vol. 35, No. 3, 2010, pp. 233-258. doi:10.1002/er.1687

10. H. Ghaebi, M. Amidpour, S. Karimkashi and O. Rezayan, "Energy, Exergy and Thermoeconomic Analysis of a Combined Cooling, Heating and Power (CCHP) System with Gas Turbine Prime Mover," International Journal of Energy Research, Vol. 35, No. 8, 2010, pp. 697-709. doi:10.1002/er.1721
11. E. Cardona, A. Piacentino and F. Cardona, "Matching Economical, Energetic and Environmental Benefits: An Analysis for Hybrid CHCP-Heat Pump Systems," Energy Conversion and Management, Vol. 47, No. 20, 2006, pp. 3530-3542. doi:10.1016/j.enconman.2006.02.027
12. P. J. Mago, N. Fumo and L. M. Chamra, "Performance Analysis of CCHP and CHP Systems Operating Follow- ing the Thermal and Electric Load," International Jour- nal of Energy Research, Vol. 33, No. 9, 2009, pp. 852-864. doi:10.1002/er.1526
13. P. J. Mago, A. Hueffed and L. M. Chamra, "A review on Energy, Economical, and Environmental Benefits of the Use of CHP Systems for Small Commercial Buildings for the North American Climate," International Journal of Energy Research, Vol. 33, No. 14, 2009, pp. 1252-1265. doi:10.1002/er.1630
14. A. A. Jalalzadeh-Azar, "A Comparison of Electrical- and Thermal-Load-Following CHP Systems," ASHRAE Transac- tions, Vol. 110, 2004, pp. 85-94.
15. A. K. Hueffed and P. J. Mago, "Influence of Prime Mover Size and Operational Strategy on the Performance of Com- bined Cooling, Heating, and Power Systems under Diffe- rent Cost Structures," Journal of Power and Energy, Vol. 224, No. 5, 2010, pp. 591-605.
16. H. Cho, S. Eksioglu, R. Luck and L. M. Chamra, "Opera- tion of Micro-CHP System Using an Optimal Energy Dispatch Algorithm," Proceedings of the ASME 2008 2nd Conference on Energy Sustainability, Jacksonville, 10-14 August 2008, pp. 747-754.
17. C. Wheeley and P. J. Mago, "A Methodology to Conduct a Combined Heating and Power System Assessment and Feasibility Study for Industrial Manufacturing Facilities," ASME International Mechanical Engineering Congress and Exposition (IMECE2011), Denver, 11-17 November 2011, Paper No. IMECE2011-62299.
18. U. S. Environmental Protection Agency Combined Heat and Power Partnership, "EPA Catalog of CHP Technolo- gies," 2008. http://www.epa.gov/chp/basic/catalog.html

Chapter 10

RECONFIGURABLE TOOLING BY USING A RECONFIGURABLE MATERIAL

Jorge Cortés, Ignacio Varela-Jiménez and Miguel Bueno-Vives

Tecnológico de Monterrey, Campus Monterrey México

INTRODUCTION

Changes in manufacturing environment are characterized by aggressive competition on a global scale and rapid changes in process technology; these require creation of production systems easily upgradable by themselves and into which new technologies and new functions can be readily integrated (Mehrabi et al, 2000). In USA; industry, government and other institutions have identified materials and manufacturing trends for 2020 (Vision 2020 Chemical Industry of The Future, 2003 & National Research Council, 1998). The materials Technology Vision committee, in the publication of "Technology Vision 2020- The U.S. Chemical Industry" has identified a number of broad goals, which are enclosed in five main areas:

- New materials
- Materials characterization
- Materials modeling and prediction
- Additives
- Recycling

An important point in this vision is the development of smart materials, which have properties of self-repair, actuate and transduce. Polymers, metals, ceramics and fluids with these special characteristics belong to this class of materials and are already used in a great diversity of applications (Vision 2020 Chemical Industry of The Future, 2003). In the other hand, the Visionary Manufacturing Challenges (National Research Council, 1998), published by the US National Academy of Sciences, presented six Grand Challenges:

- Integration of Human and Technical Resources
- Concurrent Manufacturing

- Innovative Processes
- Conversion of Information to Knowledge
- Environmental Compatibility
- Reconfigurable Enterprise

To reach these challenges, innovative processes to design and to manufacture new materials and components along with adaptable, integrated equipment, processes, and systems that can be readily reconfigured for a wide range of customer requirements or products, features, and services are needed (National Research Council, 1998). The field of smart materials and structures is emerging rapidly with technological innovations appearing in engineering materials, sensors, actuators and image processing (Kallio et al, 2003). One of the smart materials is Nickel – Titanium alloy (NiTi) that possess an interesting property by which the metal 'remembers' its original size or shape and reverts to it at a characteristic transformation temperature (Srinivasan & McFarland, 2001). Next manufacturing system generation requires of reconfigurable systems which go beyond the objective of mass, lean and flexible manufacturing systems. Because of the manufacturing trends towards a customer focused production. A reconfigurable manufacturing system is designed in order to rapid adjustment of production capacity and functionality, in response to new circumstances, by rearrangement or change of its components (Mehrabi et al, 2000). As can be seen in Fig. 1, there are many aspects of reconfiguration, such as, configuration of the product system, reconfiguration of the factory communication software, configuration of new machine controllers, building blocks and configuration of modular machines, modular processes, and modular tooling. So that, the development and implementation of key interrelated technologies to achieve the goals of reconfigurable manufacturing systems are needed.

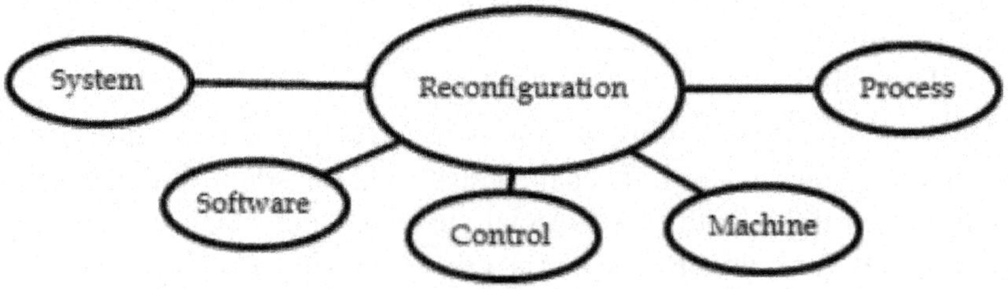

Figure 1: Aspects of reconfiguration (Mehrabi et al, 2000).

Of relevant importance are the control, monitoring and sensing of reconfigurable manufacturing systems. By noting that the system configuration changes, the parameters of the production machines and some other physical parameters will change accordingly. The controller and process monitoring systems should have the ability to reconfigure and adapt themselves to these new conditions (Mehrabi et al, 2000).

Research Justification

The use of NiTi requires proper characterization according to the environment surrounding the material when it is applied in some device; due to this requirement, a constitutive model is needed in order to relate the microstructure and thermo-mechanical behavior of the material. In the manufacture industry, a variable shape die has always been an attractive idea to reduce design time and costs, since it allows as many designs to be rapidly manufactured at nearly free cost, using one tool for several shapes (Li et al, 2008). The use of NiTi as an actuator in manufacturing systems is an opportunity area, allowing that several products can be formed by the same tool; this way, NiTi will help to evolve the traditional manufacturing industry.

Research Aim

To develop a reconfigurable manufacture system for sheet metal/plastic forming controlled by NiTi actuators and to formulate a constitutive model of its thermo- mechanical behavior.

CONSTITUTIVE MODEL

NiTi is a smart material with properties such as shape memory effect (SME) and superelasticity (Chang & Wu, 2007). SME involves the recovery of residual inelastic deformation by raising the temperature of the material above a transition temperature, whereas in superelasticity, large amounts of deformation (up to 10%) can be recovered by removing the applied loads (Azadi et al, 2007). The microscopic mechanisms involved in SME are strongly correlated to the transformation between the austenite parent phase at high temperatures and the martensite at low temperatures (Lahoz & Puértolas, 2004). It is a reversible, displacive, diffusionless, solid– solid phase transformation from a highly ordered austenite to a less ordered martensite structure (McNaney et al, 2003). Austenite has a body centered cubic lattice while martensite is monoclinic. When NiTi with martensitic structure is heated, it begins to change into the austenitic phase. This phenomenon starts at a temperature denoted by As, and is complete at a temperature denoted by Af. When austenitic NiTi is cooled, it begins to return to its martensitic structure at a temperature denoted by Ms, and the process is complete at a temperature denoted by Mf (Nemat-Nasser et al, 2006). Because austenite is usually higher in strength than martensite, a large amount of useful work accompanies the shape change. Austenite exhibits higher stiffness than martensite (De Castro et al, 2007). When NiTi is stressed at a temperature close to Af, it can display superelastic behavior. This stems from the stress-induced martensite formation, since stress can produce the martensitic phase at a temperature higher than Ms, where macroscopic deformation is accommodated by the formation of martensite. When the applied stress is released, the martensitic phase transforms back into the austenitic phase and the specimen returns back to its original shape (Nemat-Nasser et al, 2006). The stress-induced austenite-martensite transformation is effected by the formation of martensitic structures which correspond to system energy minimizers (McNaney et al, 2003) as result of the need of the crystal lattice structure to accommodate to the minimum energy state for a given temperature (Ryhänen, 1999). Shaw explains in more detail martensite behavior, affirming that due to its low degree of symmetry, the martensite exists either as a randomly twinned structure (low temperature, low stress state) or a stress-induced detwinned structure that can accommodate relatively large, reversible strains. Fig. 2 shows the thermomechanical response of a wire specimen. The specimen is first subjected to a load/unload cycle at low temperature, leaving an apparent permanent strain. The material starts in a twinned martensite (TM) state and becomes detwinned (DM) upon loading. The specimen is then subjected to a temperature increase while holding the load. The SME is seen as the strain is recovered and the material transforms to austenite (A). The temperature is then held at high value and the specimen is again subjected to a load/unload cycle. In this case the material shows superelasticity and transforms from austenite to detwinned martensite during loading and then back to austenite during unloading (Shaw, 2002).

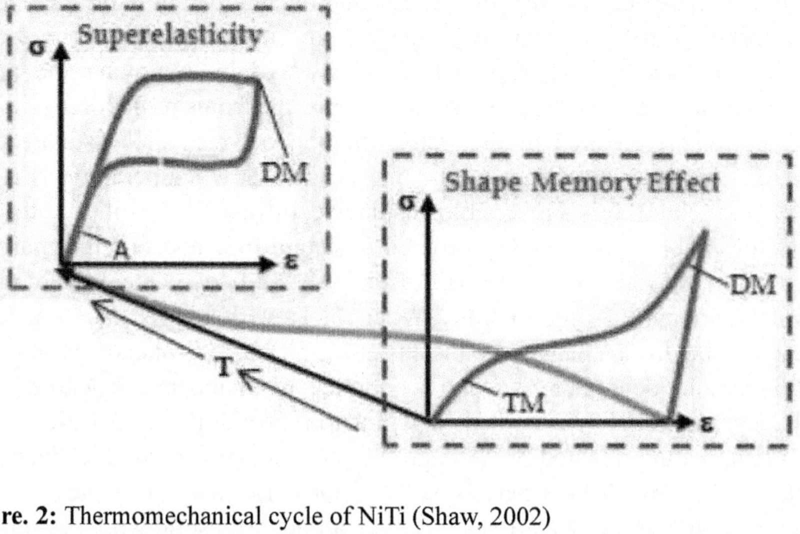

Figure. 2: Thermomechanical cycle of NiTi (Shaw, 2002)

It is considered that composition (Nemat-Nasser et al, 2006) and heat treatments have effect on the temperature at which material exhibits SME, called transformation temperatures (TTR) which are the prerequisite for the material to exhibit the SME and are one of the key parameters for SME based actuation, they also define the proper application for a certain NiTi composition alloy (Malukhin & Ehmann, 2006). Establishment of a constitutive equation for phase transformation in NiTi requires considering the Stress-Strain-Temperature behavior and the phase transformations shown in Fig. 2, from which is observed that volume fraction of each microstructure depends of the strain and temperature conditions; it also has influence on the mechanical behavior of the material. The possible phase transformations that can occur on NiTi are shown in Fig. 3, strain induces detwinned martensite while temperature increase induces austenite. As shown in Fig. 2, at low temperature and low stress, transformation of twinned martensite into detwinned martensite is started, and continues its plastic strain and then stress is released. When temperature is increased detwinned martensite transforms into austenite. If high temperature is kept and strain is applied superelasticity occurs and transformation of austenite into detwinned martensite is started, it finishes when stress is released and then microstructure transforms back into austenite.

According to Fig. 3, phase transformations on NiTi are:

1. Twinned martensite to Detwinned martensite (Strain induced)
2. Detwinned martensite to Austenite (Temperature increase induced)
3. Austenite to Detwinned Martensite (Strain induced)
4. Twinned martensite to Austenite (Temperature increase induced)
6. Austenite to Twinned Martensite (Temperature decrease induced)

Up to day, several studies have been made about NiTi, but there is a lack in the development of numerical analysis of the phenomenology of the material since its application requires a proper characterization; a constitutive model is needed in order to relate microstructure and thermo-mechanical behavior of NiTi. A similar model was developed by Cortes (Cortes et al, 1992) for determining the flow stress of aggregates with phase transformation induced by strain, stress or temperature and demonstrated its use for stainless steels. This model has also been applied on shape memory polymers (Varela et al, 2010) and it has been extended for modeling the displacement on electroactive polymers (Guzman et al, 2009) through a phase transformation approach, induced by some external stimulus. The model is based on an energy criterion which defines the energy consumed to deform the phases in the system as being equivalent to energy consumed to deform the aggregate and it is able to predict the flow stress behavior of the material. In order to apply Cortes' constitutive model on SMAs, experiments have to be carried out; the austenite, twinned martensite and detwinned martensite are considered as the aggregates and the microstructural transition between them becomes the basis of the constitutive model; this way, the constitutive expression will result in terms of the mechanical properties of each phase and its volume fraction.

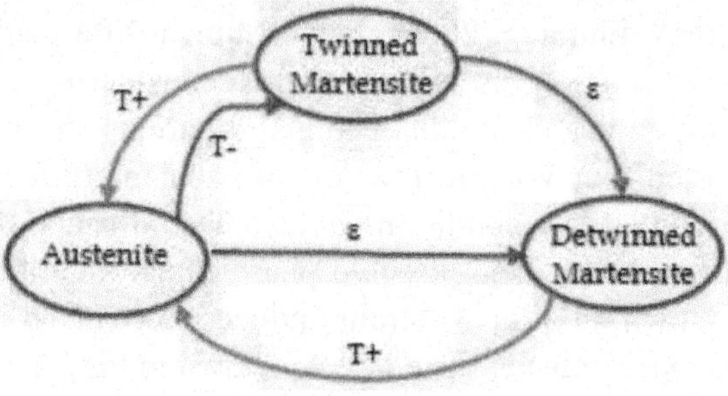

Figure. 3: Phase Transformations on NiTi and their induction stimulus.

Constitutive Model of Flow Stress

In the case of the present aggregate composed of austenite and martensite, based on Cortes model (Cortes et al, 1992), V_f of each structure or aggregate are defined as:

$$V_{fa}=\frac{V_a}{V_t} \quad V_{ftm}=\frac{V_{tm}}{V_t} \quad V_{fdm}=\frac{V_{dm}}{V_t} \tag{1}$$

where subscripts a, tm and dm indicate austenite, twinned martensite and detwinned martensite, respectively . Cortes constitutive model of flow stress for multi phases aggregate (Cortes et al, 1992) applied on NiTi is

$$\sigma_{NiTi}=V_{fa}\cdot\sigma_a+V_{ftm}\cdot\sigma_{tm}+V_{fdm}\cdot\sigma_{dm} \tag{2}$$

where σ_{NiTi} is stress of NiTi and σ_{a}, σ_{tm} and σ_{dm} are stress of each structure.

Kinetics of Strain/Temperature Induced Twinned Martensite-Detwinned Martensiteaustenite Phase Transformation

Based on the thermomechanical behavior of Fig. 2 and the phase transformations shown in Fig. 3, volume fraction of the microstructures varies by:

$$V_{fa}+V_{ftm}+V_{fdm}=1 \tag{3}$$

$$V_{fa}=\left(1-V_{fa-dm}\right)\cdot V_{fa_0}+V_{ftm-a}\cdot V_{ftm_0}+V_{fdm-a}\cdot V_{fdm_0} \tag{4}$$

$$V_{fdm} = \left(1 - V_{fdm-a}\right) \cdot V_{fdm_0} + V_{ftm-dm} \cdot V_{ftm_0} + V_{fa-dm} \cdot V_{fa_0} \tag{5}$$

where subscripts 0 indicate the initial valume of each volume fraction. For strain induced detwinned martensite phase transformation:

$$V_{fdm} = \left[1 + \left(\frac{\varepsilon}{\varepsilon_c}\right)^{-B}\right]^{-1} \tag{6}$$

For temperature induced austenite phase transformation:

$$V_{fa} = \left[1 + \left(\frac{T}{T_c}\right)^{-B}\right]^{-1} \tag{7}$$

where B is a fitting constant; while εc and Tc represent the values of strain and temperature, respectively at which 50% of the phase transformation is occurred. Experimental work is required for determining these values for each phase transformation. Substituting (6) and (7) in (4) and (5):

$$V_{f_a} = \left\{1 - \left[1 + \left(\frac{\varepsilon}{\varepsilon_{C3}}\right)^{-B_3}\right]^{-1}\right\} \cdot V_{f_{a0}} + \left[1 + \left(\frac{T}{T_{C4}}\right)^{-B_4}\right]^{-1} \cdot V_{f_{tm0}} + \left[1 + \left(\frac{T}{T_{C2}}\right)^{-B_2}\right]^{-1} \cdot V_{fdm_0} \tag{8}$$

$$V_{f_{dm}} = \left\{1 - \left[1 + \left(\frac{T}{T_{C2}}\right)^{-B_2}\right]^{-1}\right\} \cdot V_{f_{tm0}} + \left[1 + \left(\frac{\varepsilon}{\varepsilon_{C1}}\right)^{-B_1}\right]^{-1} \cdot V_{f_{tm0}} + \left[1 + \left(\frac{\varepsilon}{\varepsilon_{C3}}\right)^{-B_3}\right]^{-1} \cdot V_{fa_0} \tag{9}$$

where subscripts 1, 2, 3 and 4 represent the B, T_c or ε_c value corresponding to that phase transformation.

Stress of Microstructures

Since NiTi contains a heterogeneous microstructure under given conditions, an incremental change test to determine hardening parameters in a given structure has to be carried out. Based in Cortes work (Cortes et al, 1992) flow stress of austenite, twinned martensite and detwinned martensite is determined under isothermal conditions, by prestraining NiTi wires at a temperature at which only one microstructure exists, and then the specimens were individually deformed at a predefined temperature. The yielding point in reloading is registered as the flow stress at that temperature and those strain conditions. From this experiment equations for estimating σ_a, σ_{tm} and σ_{dm} are determined. These should be of the form:

$$\sigma_a = K_a \cdot \varepsilon^{N_a} \tag{10}$$

$$\sigma_{tm} = K_{tm} \cdot \varepsilon^{N_{tm}} \tag{11}$$

$$\sigma_{dm} = K_{dm} \cdot \varepsilon^{N_{dm}} \tag{12}$$

where K and N represent material constants which are determined experimentally. By substituting equations (3) and (8)-(12) into equation (1) stress on NiTi can be described relating thermomechanical behavior with microstructure

RECONFIGURABLE DIE

Conventional type of mold fabrication involves time and money investment to achieve design of a die; the concept about forming a die of variable shape has always been attractive as a means of rapid iterations and almost cost free (Li, 2008). Thus, multi forming methods have been developed in order to achieve reconfigurability of the process.

Multi Point Forming (MPF)

This method has been used to replace solid dies with three-dimensional surfaces. The main key of the MPF is the two matrices of punches allowing that create a three-dimensional surface which forms according to the shape of the design; this way the surface can be approximated to a continuous die, as shown in Fig. 4 (Zhong-Yi, 2002).

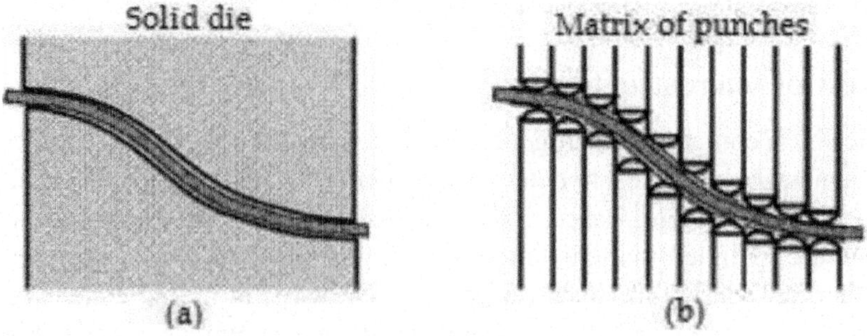

Figure. 4: (a) Conventional die forming; (b) Multi-point forming (Zhong, 2002).

MPF is based on controlling the elements and punches, hence, a matrix on punches can be shaped as required (Li, 2002). Fig. 5. illustrates the parameters related to the reconfigurable process: design and manufacture of the pin heads;

since its shape, size and length play an important role in the arrangement of the closed matrix (Walczyk, 1998). Design of a tool based on multi-point technique has many considerations since it involves several variables and many issues and problems use to occur such as dimpling, buckling and non linear deformation of the material; due to this issues four main designs have been researched with different punches types (Li, 2002).

- Multi-point full die
- Multi-point half die
- Multi-point full press
- Multi-point half press

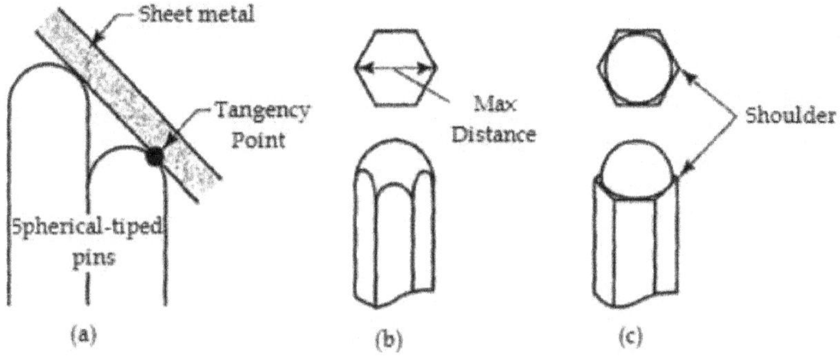

Figure. 5: Spherical and Hexagonal pin head designs (Li, 2002).

The arrangement of each design is described and shown in Table 1 and Fig. 6, respectively.

Table 1: Different types of MPF

Type of Punch	Adjustment	Required Force	Drawing / Mark
Fixed	Before Forming	Small	
Passive	While Forming	None	
Active	Free Movement	Large	

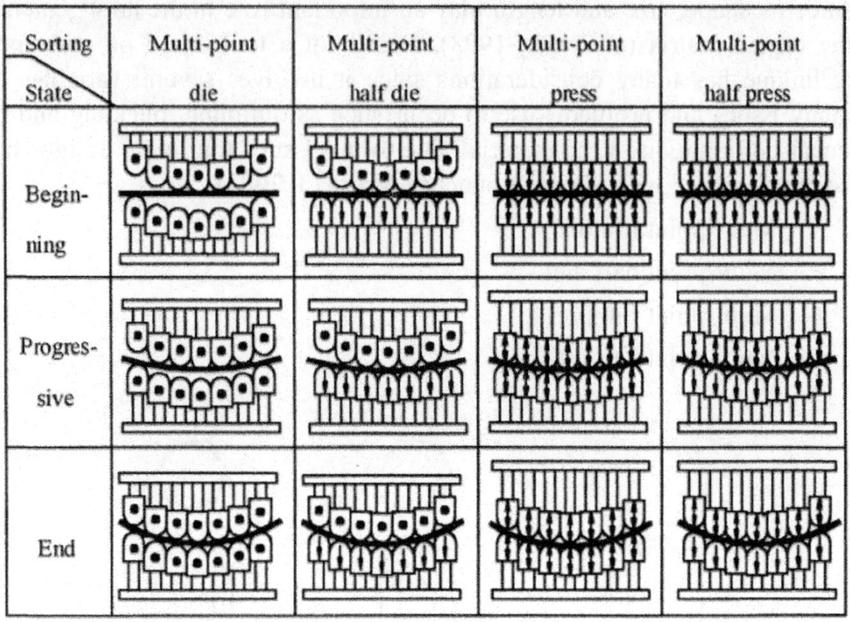

Sorting
State
Multi-point die
Multi-point half die
Multi-point press
Multi-point half press
Begin-ning
Progres-sive
End

Figure. 6: Different types of Multi-point forming and the interaction with the process (Walczyk, 1998).

Multi-Point Sandwich Forming (Mpsf)

MPSF is an accessible method to manufacture components in small batches. Fig. 7 represents the MPSF method which uses an interpolator material to assure the surface quality of the metal sheet, this last also depends of the tool elements and the position between the pins (Zhang, 2006).

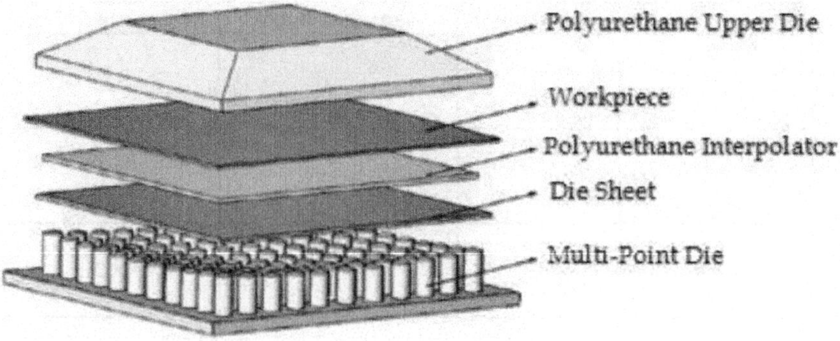

Figure. 7: Schematic Components for MPSF (Zhang, 2006).

Digitized Die Forming (DDF)

With DDF, forming procedures and integration between parameters such as deformation path, sectional forming; punches and control loop are being developed in order to avoid forming defects. The process is shown in Fig. 8 (Li, 2007).

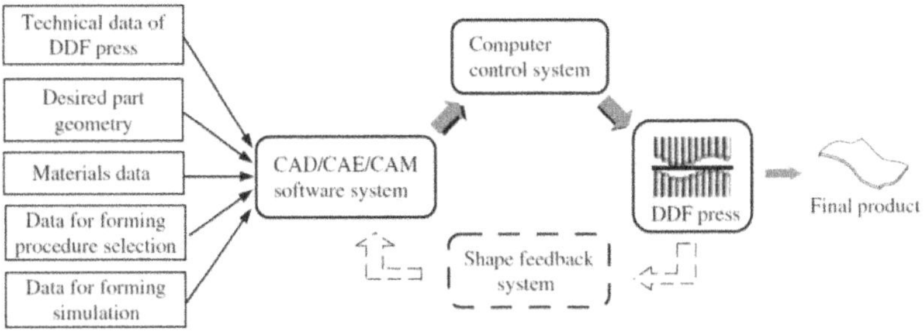

Figure. 8: Schematic of DDF integration system (Li, 2007).

Active Multi Point Forming Parameters

Pin Head

A pin head must be strong enough to support the mechanical and thermal loads of the material to manufacture avoiding the problems in the final piece such as bending, buckling and dimpling (Walczyk, 1998). In addition to normal or vertical load, the pin will be subjected to lateral load depending the height and of the adjacent pins on the shape; this variable is a key factor on the pin head design (Schwarz, 2002), as shown in Fig. 9.

Uniformity in the pin heads and elements make the design easy to fabricate and assembly in the arrangement of the matrix, a comparison of different geometries evaluating the crosssectional area shape is shown in Table 2 (Schwarz, 2002).

Table 2: Comparison of Cross-Sectional Geometry

Pin Cross-Section Shape	Equilateral Triangle	Square	Hexagonal	Circle
Number of Sides	3	4	6	6
No. Isolated straight load paths	0	2	3	0

Also the structure and the size of the pins affect the quality of the piece and it is recommended the use of square shape elements, as shown in Fig. 10, in dense packed arrangements of matrices (Schwarz, 2002).

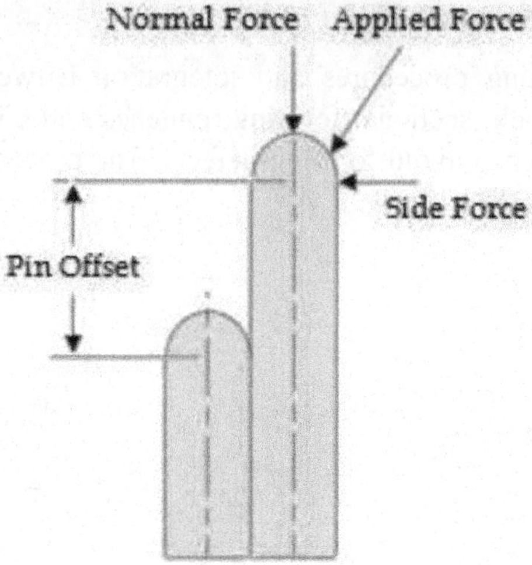

Figure. 9: Pin head forces interaction and offset.

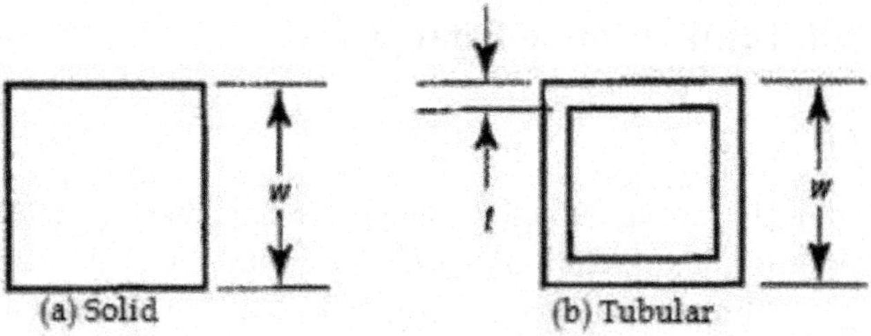

Figure. 10: Square design a) Solid and b) Tubular (Schwarz, 2002).

The use use of tubular elements is considered if the weight of the die has to be reduced, however it is not always is the best approach considering the scale and size of pins and the forces of the process (Schwarz, 2002).

Actuators

Each design of a shape has different means of independently moving pins in a large matrix arrangement (Walczyk, 2000). Automation of pin setting was not realized until in 1969 Nakajima positioned a matrix of pins controlled by a vibration mechanism mounted to a three-axis servomechanism (Nakajima, 1986). It is shown in Fig. 11. Researches from different groups such as, the Massachusetts Institute of Technology (MIT) developed a Sequential Set-up Concept, the Rensselaer Polytechnic Institute build up a Hydraulic Actuation Concept and the Northrop Grumman Group Corporation created a Shaft-driven Lead screw Concept (Walczyk, 2000)

Sequential Set-up (SSU) by MIT

Each hollow pin has a threaded nut passed from its base. The pin moves up and down as the lead screw rotates, and the next pin prevent the rotation movement; the design eliminates the matrix external clamping force to position the pin heads.

Hydraulically Actuation (HA) by Renselaer Polytechnic

Individual elements are essentially hydraulic actuators, controlled by an in-line servo valve. The hydraulic pressure makes the element rise from the initial position maintaining the height until the pressure is released.

Shaft Driven Lead Screw (SDL) by Northrop Group

This method depending on the need of the process needs a single or dual electric motor (one each on opposing sides of the die) mounted externally to drive worms mounted on cross shafts; the worm gear is connected to each pin´s lead screw.

Table 3 shows a comparison between these designs.

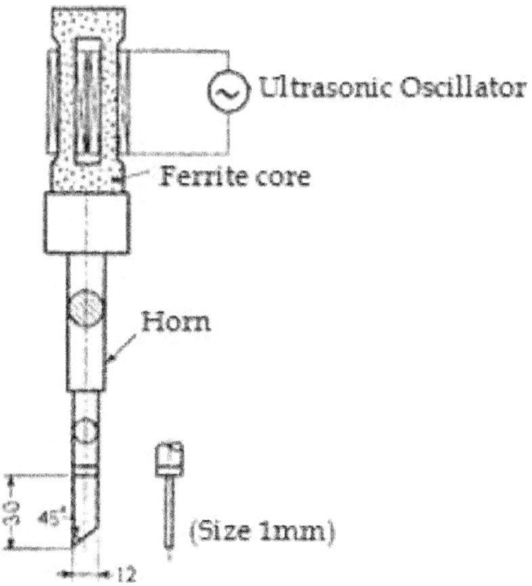

Figure. 11: Nakajima servomechanism (Nakajima, 1986).

Table 3: Comparison of Actuation schemes for Dies (Walczyk, 2000)

Characteristic	SSU	HA	SDL
Matrix of Pins	42x64 (28.6mm pin)	48x72 (25.4mm pin)	42x64 (28.6mm pin)
Number of Actuators	19 (16 drive motors, X, Y and Z axes)	1 (Hydraulic pump)	42 (drive motor per row)
Number of position control devices	0	3456 (servo valve per pin)	2688 (clutch per pin)
Number of sensors	19	1	42 (encoders per motor)
Potential mayor of positioning error	Backlash in lead screw	Insufficient platen stiffness	Rotational compliance
Setting mode	Serial	Parallel	Parallel
Potential control mode error	Lead screw is continuously engaged	Moving pins are in contact with platen	Clutch does not slip
Concept Design	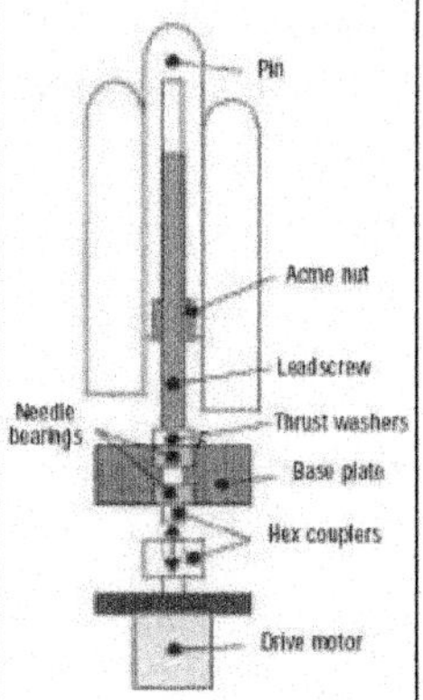	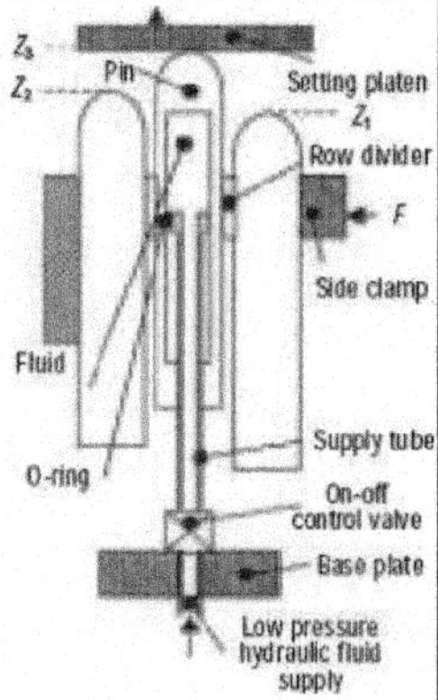	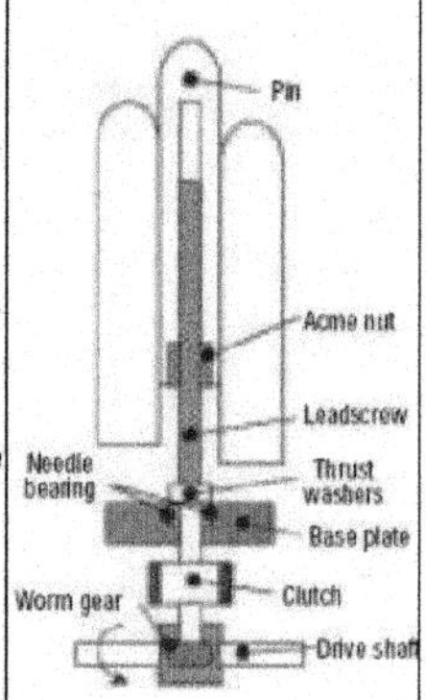

As shown in table 3; the use of common actuators have problems to form a continuous even surface, due to the size of the actuators the element diameter has to be at least 25.4mm, also the actuators and the relationship in a potential control mode error by mechanical characteristics such as fatigue and cycles limit the use of the die in order to make different shapes (Walczyk, 2000).

Matrix Arrangement

Design and manufacturability of the pins impact the position and design of the matrix due to the number of sides on the cross-sectional shape geometry. Fig. 12 shows the key factor of densely packed pin heads in a matrix; allowing management of the load and maintaining a smoother surface when subjected to loads. The contact elements have to be maximized while the gaps between elements need to be minimized (Schwarz, 2002).

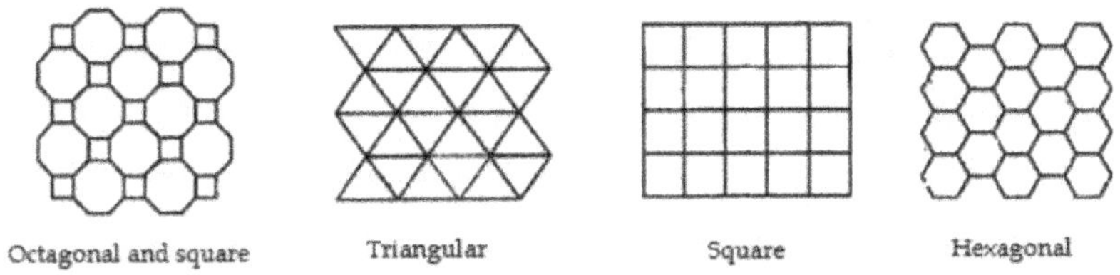

Figure. 12: Various Cross-Sectional shapes for die pins (Schwarz, 2002).

Shown in Fig. 12 is a discrete digitalized arrangement to a continuous surface with multi point forming die technology. It is important to have as many pins as possible since a poor transition surface may result in delicate wrinkling defect and possible cracking of the work piece (Peng, 2006).

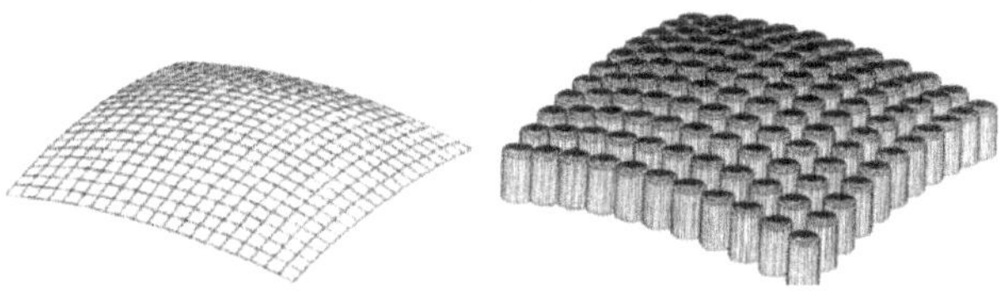

Figure. 13: Discrete approximation to a continuous surface square position (Rao, 2002).

Control and Software

The desired part to be manufactured generating a controlling data in height of the piece is sent to a control system to perform the DDF (Li, 2007). Fig. 15. shows the different technologies merging to make the software design and control available to a reconfigurable design. An open loop control system and electronic on a single pin element, are used to evaluate a simpler circuit and timed software to excite the actuator (Walczyk & Hardt, 1998), as shown in Fig. 14.

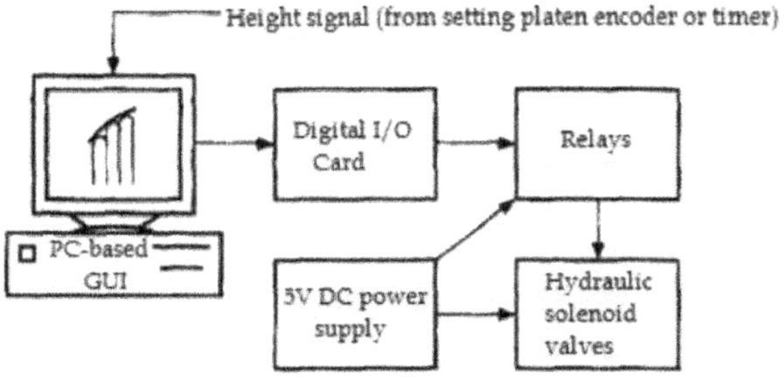

Figure. 14: Test Schematic of Die Control System (Walczyk & Hardt, 1998).

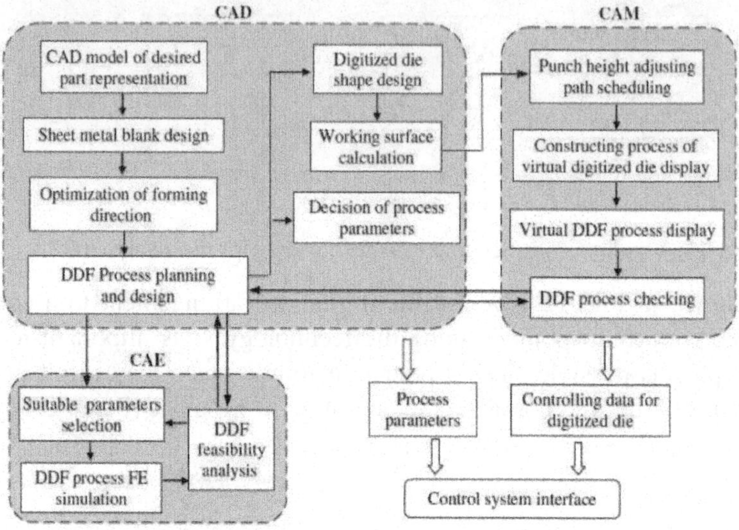

Figure. 15: CAD/CAE/CAM software control for DFF (Li, 2007).

Shape Memory Alloy Actuators

The uses of SMA as actuators nowadays are not only medical devices; there are also novelty actuators such as (Humbeeck, 1999):

- Fashion and gadgets: single products were created from cell phone antennas, eye glasses frames and in the clothing industries frames for brassieres and wedding dresses pericoats (Duering, 1990).
- Couples: Heat-recoverable couplings of an F-14 hydraulic turbine were the first large scale produce actuator (Duering, 1990).
- Micro-actuators: The central Research Institute of Electric Power Industry in Japan built a piston-driver from a 2mm diameter wire based on 26 bars, having a life cycle over 500 000 cycles minimum.
- Adaptive materials: A vibrator frequency control of a polymer beam has been used to increase the natural frequency of the composite beam.
- Other applications: Wear cavitations defects where hydraulic machinery are used like water turbines, ship propellers and sluice channels (Jardine et al, 1994).

Development of Reconfigurable Die Based On Niti Actuator

An active multi point forming tool, based on NiTi wires as main actuators is developed. The devices is known as 'reconfigurable die'.

Hypothesis

The main issue on the development of multi point surfaces is that a high density matrix of pins is required, as smaller are the pins smoother is the surface that can be formed. This issue can be solved by using a small actuator that allows formation of a dense pins matrix. SME of NiTi can be applied to achieve the movement of a mechanism that controlls the movement of each pin

Methodology

Development of a reconfigurable die follows the concept of DDF. Hence, it is required to design a mechanism, a controller and a graphical user interface (GUI), the full system is shown in Fig. 16.

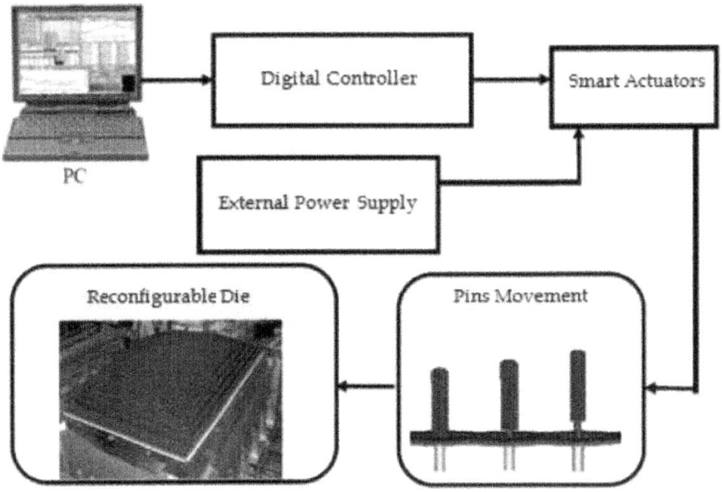

Figure. 16: Overall Process Variables.

Mechanism Design

A design of a reconfigurable die proposed has been reviewed in order to identify its components and characterize them, sucha as the length of the shafts, springs parameters and SMA properties (length, diameter, electric current).

Functional Prototype Description

Fig 18 shows the mechanism that controls the vertical movement of each square pin. Each pin has a SMA wire subjected to a spring that deforms it, thus, NiTi has a martensitic structure. When a electric pulse activates the electric current the wire will be heated reaching the austenitic structure returning the wire to the non deformed position pushing the springs and the shaft, that causes that the pin rotates and move up; when the pulse is inactive the springs will deforms the SMA again and the cylce is restarted.

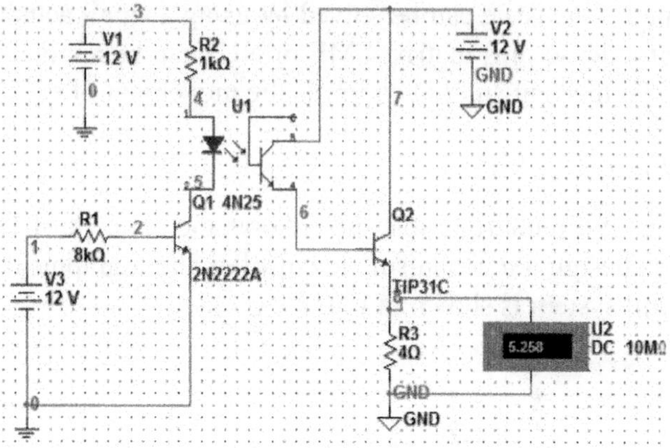

Figure 17: Electronic design. Power and control circuits, R3 represents NiTi wire.

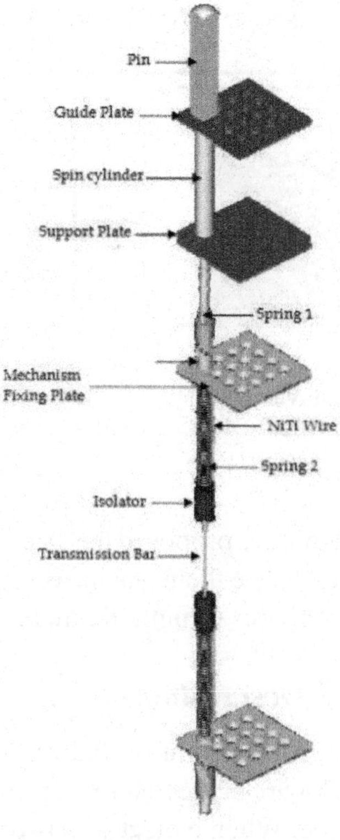

Figure 18: General Design of the reconfigurable die.

Circuitry

The design consist in a basic control and power electronic circuitry which supplies the current needed to the SMA. Circuit is illustrated in Fig. 17.

Software

It was programmed with Labview by National Instrument and consists of three set of parts. The first one is the image codification from a solid figure to a virtual 3D figure. The second part is the pin head elements array and virtual configuration in the software; finally the last part is the digital output of electric pulses needed per pin head to change its height. Image codification is performed by pictures taken from the object of interest, in order to make a virtual solid image in a 40 by 40 matrix. The number of views of the object depends of the geometry, a maximum of three pictures can be uploaded (sideway, front, and top views) in order to arrange the pins matrix. The file from the generated matrix is then placed in the next file, resulting of a specific height for each element; this is then saved in a file as an array in order to visualize the pin design and making a blank. Each element has a resolution of a 28.5 pixel per pin. Fig 19 shows the Human Machine Interface (HMI) software and Fig. 20 shows the matrix of the pins height loaded on the software.

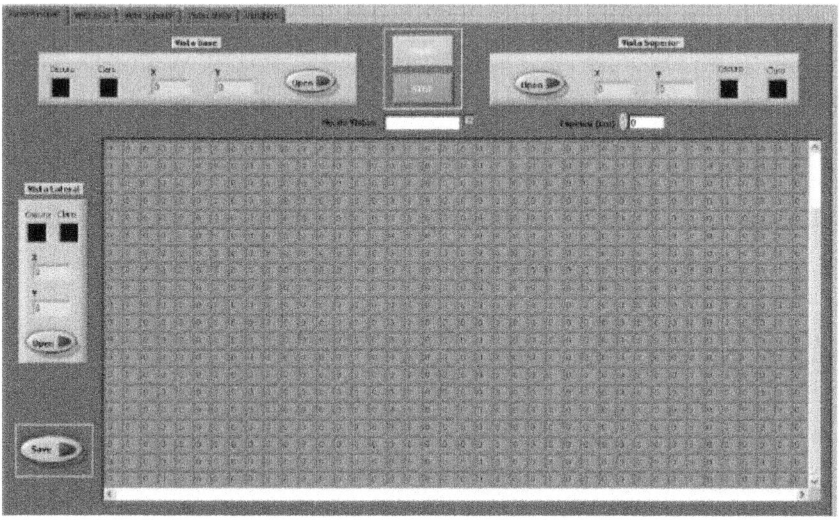

Figure 19: Software HMI.

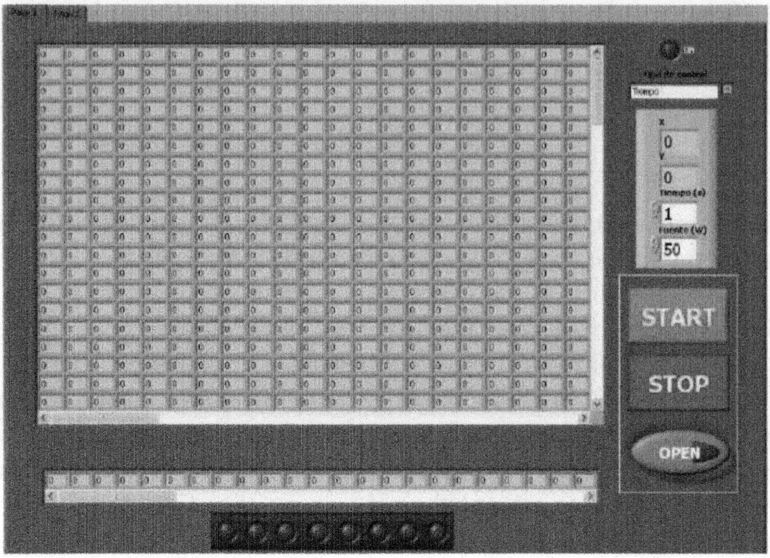

Figure 20: Matrix pin height.

Control

Digital pulses: A digital pulse, as shown in Fig. 21, consists in a square wave of direct current output, on which the duty cycle is fixed, resulting a 50% high and 50% low. In this case the "high time" the actuator will be activated with current and the "low time" the signal will be deactivated in order to cool down the actuator.

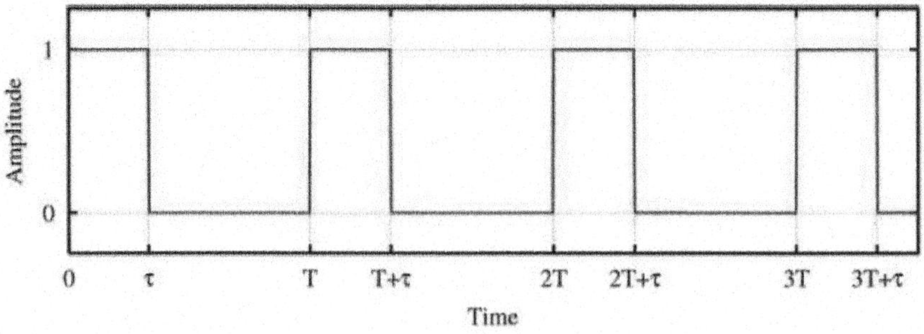

Figure: 21: Digital pulse.

The Duty Cycle (DC) represents the pulse duration divided by the pulse period, where is the duration that the function is actve hight (normally when the voltage is greater then zero) and T is the period of the function.

Pulse-width Modulation (PWM): is one of the most efficient ways to provide electrical power between the ranges fully on and fully off. It is a great electric tool to supply voltage/current in ht power electronics field to devices such as electric stoves, robot sensors, dimmers. The main characteristics are the variation and switching between the high and low levels in high frequencies ranges. PWM, as shown in Fig. 22, uses a rectangular pulse wave whose pulse width is modulated resulting in the variation of the average value of the waveform. The wave has a changing duty cycle, making many pulses in a period of desired time.

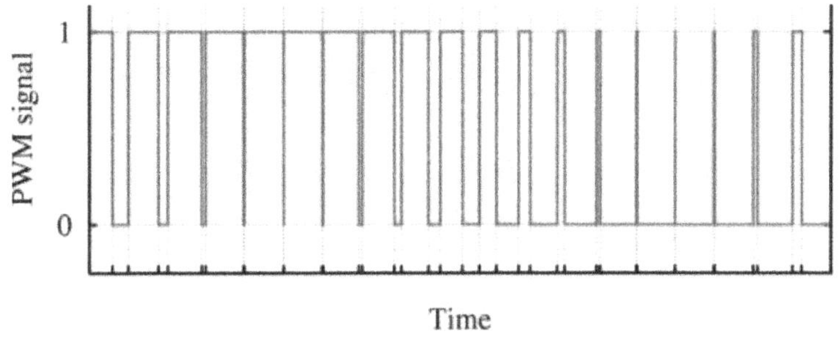

Figure 22: PWM signal.

In order to select the most appropriate output design in the system, a Design of Experiments (DOE) it is implemented, it is shown in Table 4.

Table 4: DOE for evaluating the electric pulse output of each pulse

DOE		
Factor	**High level**	**Low level**
Time of pulse	4 second	2 second
Wave output	Single Pulse	PWM Pulse
Power Type	5.5 Watts	13 Watts

For testing purposes, the PWM has a high frequency of 10 KHz, for the single pulse the duty cycle is a 50%. The time of pulse is the total amount of the pulse. According to the tests, the results indicates that the factor of type of power and time of pulse have an impact and the best results on the actuator to return to the original position and having a bigger recovery force is the single pulse in a lower time with a medium range power supply. The control design to activate the movement of the elements, there are two different methods that can be considered, as described in Table 5.

Table 5: Pin actuation scheme

Method	Definition	Elapsed time
Serial	Elevates the height of each pin from each row is elevated one at a time.	Longer
Parallel	Elevates the height of all pins from each row is elevated.	Shorter

The difference in time setting can be calculated simply by adding the quantity of total cycles from all the pins in the matrix and multiplying by the total time per one pulse in seconds.

Final Prototype

The final alpha prototype is shown in Figs 23 and 24.

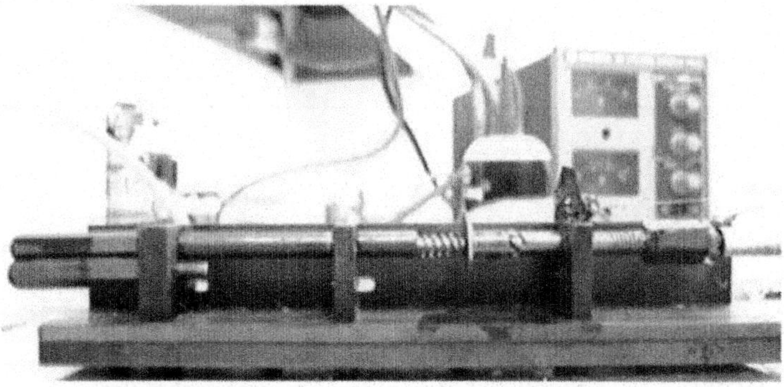

Figure. 23: Machined prototype.

Figure. 24: Pin element changes height.

Operation Process

The manufacture process and the use of this tool, allows a technological advantage in the design of a multi-point shape with a reconfigurable die. Its use involves CAD, CAM and CAE technologies as shown in Fig. 25.

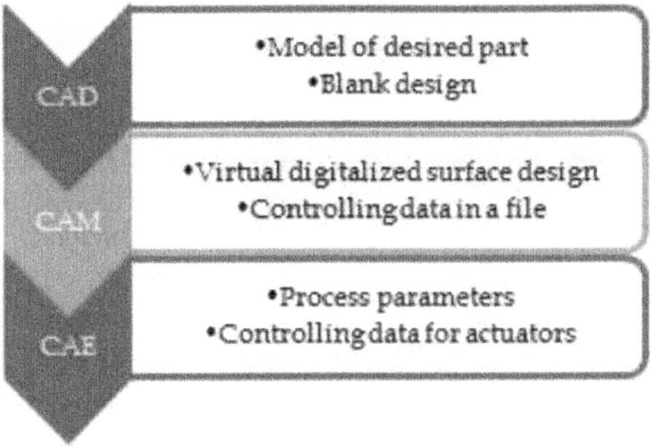

Figure. 25: Technologies used.

Image: the image depends on the views taken by the camera and the position of the object, the files are imported to the matrix arrangement software file to create a virtual object. Virtual Matrix: when selecting the geometry of the matrix, the columns and rows are selected to visualize the object. The object then will be saved in a matrix file in the hard drive, containing the position of the elements needed from all the matrix, this allow the user to save as many designs and objects without repeating the image step every time. Physical Matrix: Importing the matrix file, then the software will calculate the number of total cycles per element needed to reach the certain height. In this file the type of control (Parallel or Serial) is defined with the Data Adquisition Card (DAQ) digital outputs. These proceeses are resumed in Fig. 26.

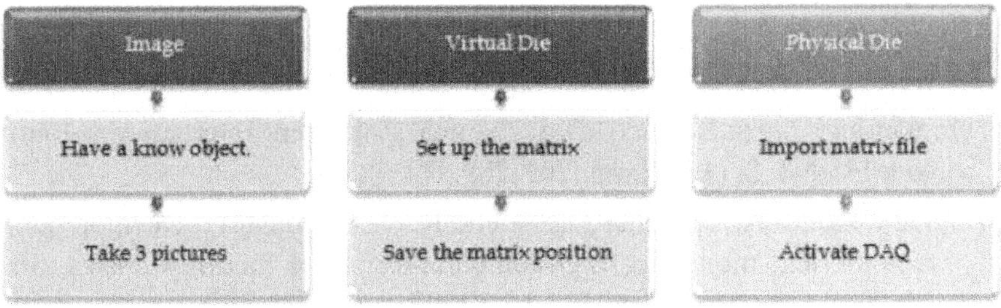

Figure. 26: Processes description.

CONCLUSIONS

The proposed cconstitutive model for stress on NiTi relates microstructure with thermomechanical behavior of NiTi. A single expression considers the 3 possible existent microstructures and their strain/temperature induced phase transformation. The size and design of a reconfigurable tool has a strong relationship with size and design of the actuator on which the elements will be positioned in a matrix array. The use of a reconfigurable actuator such as SMA, makes a more detail design of the pieces and decrements the size and shape of the overall elements, as seen previously in this chapter, making a highly dense pin head per matrix area resulting on a more continuous shape for the discrete shape. The use of a step by step method such as the one proposed, makes the process an adaptable and enhanced the Vision of Manufacturing Challenges 2020 a reachable goal. The establishment of the mechanical and electronic parameters of the proposed in order to make a functional prototype, demonstrates that the use of shape memory alloy as actuator can be possible. The methodology recommended complies the flexibility of a reconfigurable tool according to the Manufacture Vision Challenges 2020, making a modular and adaptable process. (From the object processing image to a virtual matrix array visualizing the final surface created by the matrix arrangement of the pin elements.)

REFERENCES

1. Azadi, B., Rajapakse, R., & Maijer D. Multi-dimensional constitutive modeling of SMA during unstable pseudoelastic behavior. International journal of solids and structures,Vol. 44, No, 20, (2007), pp. 6473-6490, ISSN 0020-7683
2. Chang, S., & Wu S. Internal friction of R-phase and B19' martensite in equiatomic TiNi shape memory alloy under isothermal conditions. Journal of Alloys and Compounds, Vol. 437, (2007), pp. 120-126, ISSN 0925-8388
3. Cortes, J., Tsuta, T., Mitani, Y., & Osakada, K. Flow stress and phase transformation analyses in the austenitic stainless steels under cold working. Japan Society of Mechanical Engineers International Journal, Vol. I35 No. 2, (1992), pp. 201-209.
4. De Castro J A., Melcher K., Noebe R., & Gaydosh D. Development of a numerical model for high-temperature shape memory alloys. Smart Materials and Structures. Vol 16. (2007). pp. 2080-2090.
5. Duering, TW., Melton, K., & Stockel D.Engineering Aspects of Shape Memory Alloys. Materials and Manufacturing Processes, (1990).

6. Guzman, J., Cortes, J., Fuentes, A., Kobayashi, T., & Hoshina, Y. Modeling the displacement in three-layer electroactive polymers using different counter-ions by a phase transformation approach. Journal of Applied Polymer Science, Vol. 112, No.6,(June 2009), pp. 3284–3293. ISSN 1097-4628
7. Humbeeck, J. Non-medical applications of shape memory alloys. Materials Science and Engineering, (1999).
8. Kallio, M., Lahtinen, R., & Koskinen, J. (2003). Smart materials. Smart materials and structures. VTT Research Program 2000-2002 Seminar. Retrieved from <http://www.vtt.fi/inf/pdf/symposiums/2003/S225.pdf>
9. Lahoz, R., & Puértolas, J. Training and two-way shape memory in NiTi alloys: influence on thermal parameters. Journal of Alloys and Compounds, Vol. 381, (2004), pp. 130–136,ISSN 0925-8388
10. Li, M. Multi-point forming technology for sheet metal. Journal of Materials Processing Technology, (2002), pp. 333-338.
11. Li, M. Manufacturing of sheet metal parts based on digitized-die.Robotics and ComputerIntegrated Manufacturing, (2007), pp. 107-115.
12. Li, M., Cai, Z., Sui, Z., & Li, X. Principle and applications of multi-point matched-die forming for sheet metal. Proceedings of the Institution of Mechanical Engineers, Part B: Journal of Engineering Manufacture, Vol. 222, (May, 2008), pp. 581-589.
13. Malukhin K & Ehmann K. Material Characterization of NiTi Based Memory Alloys Fabricated by the Laser Direct Metal Deposition Process. Journal of Manufacturing Science and Engineering. Vol. 128, (2006), pp. 691-696, ISSN 1087-1357
14. McNaney J., Imbeni V., Jung Y., Papadopoulos P., & Ritchie R. An experimental study of the superelastic effect in a shape-memory Nitinol alloy under biaxial loading.
15. Mechanics of Materials, Vol. 35 (2003), pp. 969–986, ISSN 0167-6636
16. Mehrabi, M., Ulsoy, G., & Koren Y. Reconfigurable Manufacturing Systems: Key to Future Manufacturing. Journal of Intelligent Manufacturing, Vol.11, No.4, (August 2000), pp. 403-419, ISSN 0956-5515
17. Nakajima, N. A Newly Develop Technique to Fabricate Complicated Dies and Electrodes with Wires. Japan Society of Mechanical Engineers International Journal, (1986).
18. National Research Council. Visionary Manufacturing Challenges For 2020, (1998). National Academies Press. Retrieved from <www.nap.edu/

readingroom/books/visionary>

19. Nemat-Nasser S., & Guo W. Superelastic and cyclic response of NiTi SMA at various strain rates and temperatures Mechanics of Materials. Vol 38, (2006), pp. 463–474, ISSN 0167-6636
20. Peng, L. Transition surface design for blank holder in multi-point forming. International Journal of Machine Tools & Manufacture, (2006), pp.1336-1342.
21. Rao, P. A flexible surface tooling for sheet-forming processes: conceptual studies and numerical simulation. Journal of Materials Processing Technology, (2002).
22. Ryhänen J. Biocompatibility Evaluation Of Nickel-Titanium Shape Memory Metal Alloy. University of Oulu. (1999). Retrieved from <http://herkules.oulu.fi/isbn9514252217/html>
23. Schwarz, R. Design and Test of a Reconfigurable Forming Die. Journal of Manufacturing Processes, (2002), pp. 77-85.
24. Shaw, J. A thermomechanical model for a 1-D shape memory alloy wire with propagating instabilities. International Journal of Solids and Structures, Vol. 39, (2002), pp.1275–1305, ISSN 0020-7683
25. Srinivasan, A., & McFarland D. Smart Structures Analysis and Design (2001), Cambridge University Press. ISBN 0-521-65026-7, Cambridge UK
26. Varela, M., Cortes, J., & Chen Y. Constitutive Model for Glassy – Active Phase Transformation on Shape Memory Polymers considering Small Deformations. Journal of Materials Science and Engineering, Vol. 4, No. 5 (May 2010), pp 14-22, ISSN 1934-8959
27. Vision 2020 Chemical Industry of The Future. Roadmap for Process Equipment Materials Technology, (2003). Retrieved from <www.chemicalvision2020.org/techroadmaps.html>
28. Walczyk, D., & Hardt, D. Design and Analysis of Reconfigurable Discrete Dies for Sheet Metal Forming. Journal of Manufacturing Systems, Vol.17, No.6, (1998)
29. Walczyk, D. A Comparison of Pin Actuation Schemes for Large-Scale Discrete Dies. Journal of Manufacturing Processes, (2000), pp. 247-257.
30. Zhang, Q. Numerical Simulation of deformation in multi-point sandwich forming. Machine Tools & Manufacturing. (2006): pp. 699-707.
31. Zhong-Yi, C. Multi-point forming of three-dimensional sheet metal and the control of the forming process. International Journal of Pressure Vessels and Piping. (2002): pp. 289-296.

Chapter 11

TOWARDS ADAPTIVE MANUFACTURING SYSTEMS - KNOWLEDGE AND KNOWLEDGE MANAGEMENT SYSTEMS

Minna Lanz, Eeva Jarvenpaa, Fernando Garcia, Pasi Luostarinen and Reijo Tuokko

Tampere University of Technology Finland

INTRODUCTION

Today the changes in the environment are those business related or manufacturing, are both frequent and rapid. Industry has talked about the adaptation to meet the changes over a decade. Adaptation as a word has gained quite a reputation. Adaptation is expected in design of products and processes and in the realization of processes. The adaptation in the field of manufacturing sector is commonly understood as operational flexibility and reaction speed to the changes and/or opportunities. However, in order to achieve the required level of adaptability a company must be able to learn. Learning is achieved through gaining and understanding feedback of a change: its quantity and direction. Gaining and understanding the feedback a company must be able to compare the past status to the new status of actions. Unfortunately, the knowledge of neither the past nor the present is in computer interpretable and comparable form. Thus, the achieved and/or imagined flexibility is slightly above non-existent in reality.

This chapter discusses the possibilities of a modular and more transparent knowledge1 management concept that provides means for representing and capturing needed information as feasible as possible while understanding that it is also the software systems that need to adapt to the changes along the physical production systems. The research approach discussed here aims to introduce new ideas for the company's knowledge management and process control by facilitating the move from technology based solutions to configurable systems and processes where the digital models and modular knowledge management systems can be configured based on needs - not based on closed legacy systems. The case implementation chosen here to illustrate this approach divides the knowledge management system into three separate layers: data storing system, semantic operation logic (the knowledge representation) and services that utilize the commonly available knowledge. The modular approach in ICT allows also software vendors to enhance their production to be more modular and configurable thus allowing the service oriented operation model to be realized. Once the storing method is separated from the logic and services, the new concepts can emerge. It is seen also that the vendors can make new business opportunities based on modular system solutions and configuration of those instead of highly tailored solutions which cannot be re-used later on.

The chapter is structured as following: the section 2 will illustrate the challenges industrial world is facing today. Section 3 summarizes the state of the art in field of knowledge modeling. Section 4 outlines needs for modularity in systems and introduces one possible solution candidate. Section 5 introduces a case implementation. Section 6 concludes the chapter and section 7 discusses about the challenges and future trends.

SET OF CHALLENGES FOR THE NEW DECADE

From simple to complex operation environment

For society to sustain and prosper, it needs along with societal, structural and organizational values a steady flow of income. For most of societies manufacturing has been and still is one of the biggest source of income. However, global competition has changed the nature of European manufacturing paradigms in past decades, see Figure 1. A turbulent production environment, short product life-cycles, and frequent introduction of new products require more adaptive systems that can rapidly respond to required changes whether or not the changes are based on product design changes or changes in the production itself. However, the technological leap in the mid-20th century provided the means to venture towards more capable systems with very highly performing components. Today the acute problem is to take full advantage of their specific capabilities. These new systems, called complex systems, are no longer reducible to simple systems like complicated ones described by Descartes, Cotsaftis (2009). Technical developments in recent years have produced stand-alone systems where high performance is routinely reached. This solid background has allowed the extension of these systems into networks of components, which are combined from very heterogenous elements, each in charge of only a part of the holistic action of the system. As the systems are process oriented instead of knowledge oriented systems, the interaction between tasks cannot be modeled, thus the effect of single interactions and relationships cannot be represented in the full systems scale.

The types of interactions are changing into a complex network of possibilities within certain limits instead of a steady and predefined process flow. This situation is relatively new and causes pressures to define the role of intended interaction. According to Chavalarias et al (2006), there is no doubt that one of the main characteristics of complex and adaptive production platforms in the future will be the ever- increasing utilization of ICT. However, while the industrial world has seen the possible advantages, the implementations fall short as a result of the required changes to the whole production paradigm, going from preplanned hierarchical systems to adaptive and self-organizing complex systems, Chavalarias et al (2006) and Cotsaftis (2009).

Chavalarias et al (2006) stated that complex systems are described as the new scientific frontier which has been advancing in the past decades with the advance of modern technology and the increasing interest towards natural systems' behavior. The main idea of the science in complex systems is to develop through a constant process of reconstructing models from constantly improving data. The characteristics of multiple-component systems are to evolve and adapt due to internal and external dynamic interactions. The system keeps becoming a different system. Simultaneously, the connection between the system and its surroundings evolves as well. When multiple-component system is manipulated it reacts via feedback, with the manipulator and complex system inevitably becoming entangled.

Paradigm	Craft Production	Mass Production	Flexible Production	Mass Customization and Personalization	Open Complex and Adaptive Production Systems
Paradigm started	~1850	1913	~1980	2000	2020
Society needs	Customized Products	Low cost products	Variety of products	Customized products	Customized on-demand products
Market	Very samll volume per product	Steady demand	Smaller volume per product	Global manufacturing and fluctuating demand	Global manufacturing and fluctuating demand
Business Model	Pull Sell-design-make-assemble	Push design-make-assemble-sell	Push-Pull design-make-sell-assemble	Pull design-sell-make-assemble	Pull design-sell-make-assemble
Technology Enabler	Electricity	Interchangeable parts	Computers	Information technology	Information and communication technology
Process Enabler	Machine tools	Moving assembly line	Flexible Manufacturing Systems, robots	Reconfigurable Manufacturing System	Self-organizing agents

Figure 1: Paradigm shift, adapted and modified from ManuFuture Roadmap published by European Commission (2003)

In complex systems, reconstruction is searching for a model that can be programmed as a computer simulation that reproduces the observed data 'well'. The ideal of predicting the multi-level dynamics of complex systems can only be done in terms of probability distributions, i.e. under non-deterministic formalisms. An important challenge is, contrary to classical systems studies, the great difficulty in predicting the future behavior from the initial state as by their possible interactions between system components is shielding their specific individual features. In this sense, reconstruction is the inverse problem of simulation. This naturally indicates that the complex system cannot be understood as deterministic system, since the predictions from Complex Systems Science do not say what will happen, but what can happen, Valckenaers et al (1994), Chavalarias et al (2006), Cotsaftis (2009) and Lanz (2010).

In general, complex systems have many autonomous units (holons, agents, actors, individuals) with adaptive capabilities (evolution, learning, etc), and show important emergent phenomena that cannot be derived in any simple way from knowledge of their components alone. Yet one of the greatest challenges in building a science of such systems is precisely to understand this link - how micro level properties determine or at least influence properties on the macro level. The current lack of understanding presents a huge obstacle in designing systems with specified behavior regarding interactions and adaptive features, so as to achieve a targeted behavior from the whole, Chavalarias et al (2006).

Due to the complexity of the system behavior and the lack of tangible and implementable research results on how complex systems theory can bring revenue to a company; implementations at the moment are scarce and acceptance varies. In order to meet the new requirements set by the evolving environment several new manufacturing paradigms have been introduced, which follow characteristics of natural systems. These paradigms are:

- Bionic Manufacturing System (BMS): The BMS investigates biological systems and proposes concepts for future manufacturing systems. A biological system includes autonomous and spontaneous behavior and social harmony within hierarchically ordered relationships. Cells as an example are basic units, which comprises all other parts of a biological system and can have different capabilities from each other, and are capable of multiple operations. In such structures, each layer in the hierarchy supports and is supported by the adjacent layers. The components, including the part, communicate and inform each other of the decisions, Tharumarajah et al. (1996) and Ueda et al. (1997).
- Fractal Factory (FF): The concept of a fractal factory proposes a manufacturing company composed of small components or fractal entities. These entities can be described by specific internal features of the fractals. The first feature is self-organization that implies freedom for

the fractals in organizing and executing tasks. The fractal components can choose their own methods of problem solving including self-optimization that takes care of process improvements. The second feature is dynamics where the fractals can adapt to influences from the environment without a formal organization structure. The third feature is self-similarity understood as similarity of goals among the fractals to conform the objectives in each unit Tharumarajah et al. (1996) .

- Holonic Manufacturing System (HMS): The core of HMS is derived from the principles behind the term 'holon'. The term holon means something that is at the same time a whole and a part of some greater whole Koestler (1968). The model of integrated manufacturing systems consists of manufacturing system entities and related domains, the structure of individual manufacturing entities, and the structuring levels of the entities. A manufacturing system is, at the same time, part of a bigger system and a system consisting of subsystems. Each of the entities posses self-description and capability for self-organization and communication, Valckenaers et al (1994) and Salminen et al (2009).

The meaning of knowledge

It is said that the world is surrounded by knowledge. Knowledge is saved into knowledge-bases and managed by knowledge management systems is something what has been stated over and over again. However, today, no matter what the vendor flyers express with colorful pictures and highly illustrative arrows, knowledge - as computers can understand it and reason with it - is not saved. The majority of the research and design effort is never captured or re-used. The interpretation of, for example a technical drawing is entirely based on the human perception and this perception may vary. "The meaning of knowledge is not captured and therefore not utilized as it has been intended."

The need today is the capability for rapid adaptation to the changes in environment based on the previously acquired knowledge. However, the challenge is precisely the input knowledge or to be more accurate: the lack of it. In a large-scale company there can be up to hundreds of different design support systems, versions, and ad-hoc applications, which are used to create the information of the current product, process, and/or production systems. The majority of systems is using proprietary data structures and vaguely described semantics. This leads to challenges in information sharing since none of those are truly able to share data beyond geometrical visualizations. The design knowledge - the design intention - if even created, remains locked inside the authoring system, Ray (2004), Lanz (2010), Jarvenpaa et al. (2010), Lohse (2006) and Iria (2009).

THE STATE OF THE ART

Data modeling

As product, process and manufacturing system design have become more and more knowledge-intensive and collaborative, the need for computational frameworks to support much needed interoperability is critical. Academia and industrial world together have provided multiple different standards for product, process and resource models ranging from conceptual models to very formal representations. However, there are some serious shortcomings in the current representations:

- Firstly, none of these can represent the needs of the industry, not even industrial sector as whole.
- Secondly these standards do not form knowledge architecture due to the missing critical parts (such as life-cycle information of products, processes and factory systems, history of past events and occurances).
- Thirdly, there does not exist a study that would outline the overlapping between these standards, Lanz et al. (2010).

Table 1 summarizes several different languages to represent data models that exist today. The list is not complete, nor it is intend to be, but it will summarize examples of standards, de facto standards and other models that are used today by industry and academia. There have been three main approaches used to create a knowledge exchange infrastructure. They are a "point-to-point" customized solution, where dedicated interfaces are created between the design tools; a "one size fits all" solution decided by the original equipment manufacturer (OEM)'s proprietary interface for design and planning and knowledge exchange between parties; and the third solution is the a "neutral and open reference architecture" based on published standards. The first approach is expensive and time-consuming for the OEM, while the second option is very cost-efficient for the OEM, but expensive for partners who are working with several OEMs. The third option has never been fully implemented, Ray (2004), Lanz (2010) and Lohse (2006).

Knowledge capture

Second large problem area is the knowledge capturing. Currently there are very few systems that can be called knowledge capturing systems. By the definition information becomes knowledge, once other parties exist, which can understand the meaning of the information and can use it for their own purposes. In large scale organizations, data regarding activities and tasks are routinely stored in an unstructured manner, in the form of images and natural language used in e-mails, word-processed documents, spreadsheets and presentations. Over time, large unstructured data repositories are formed, which preserve valuable information for the organization, if this information can ever be found or used.

Table 1: Means for representing domain knowledge

LANGUAGE/STANDARD	USED IN PROJECTS AND STANDARDS	DESCRIPTION AND USE
EXPRESS	Standard for the Exchange of Product model data (ISO 10303 STEP), Open Assembly Model (OAM), Core Product Model (CPM), Krima et al (2009)	Defining the connections between the artifacts
CommonCADS	EUPASS Ontology, Lohse (2006)	Definition of interdependencies between classes
Web Ontology Language, Description Logics (OWL DL)/ Resource Description Framework (RDF)	Core Ontology Lanz (2010), ontoSTEP, Krima et al (2009)	Definition of interdependencies between classes and artifacts
First Order Logic (FOL)	Core Ontology, Lanz (2010)	Definition of interdependencies between classes
Common Logic Interchange Format (CLIF)	Process Specification language (PSL)	Describing what actually happens when a process specification executes and for writing constraints on processes, Bock & Gruninger (2005).
eXtensive Mark-up Language (XML)	Core Manufacturing Simulation Data (CMSD)	Used for the exchange manufacturing resource data
Automation ML	Knowledge Integration Framework ROSETTA (2010)	Representation Language of entities ROSETTA (2010)
Pabadis Promise Product and Production Process Description Language (P5DL)	OWL based language in FP6 Pabadis'Promise (2006)	P5DL used for description of products (as STEP) with their commercial and control relevant data and their necessary control applications and description of manufacturing processes with their hosting resources and necessary control functions, FP6 Pabadis'Promise (2006).

Thus, a challenging research issue is to consider how information and knowledge is spread across numerous sources, and how it can be captured and retrieved in an efficient manner. Unfortunately, traditional information retrieval

(IR) techniques not only tend to underperform on the kinds of domain-specific queries that are typically issued against these unstructured repositories, but they are also often inadequate, Iria (2009). The capturing of knowledge should start already from the creation of knowledge, where the engineer knows the meaning of the models and documents he/she is creating. This meaning should be captured in a form of computer readable format, such as a formal ontology, for further use.

Knowledge and meaning

According to DoHS (2008) increasing trend can be found from ongoing research in different domain contexts on using emerging technologies such as ontologies, semantics and semantic web (Web 2.0), to support the collaboration and interoperability. In recent years there have been a lot of activities concerning the domain and upper ontologies for manufacturing. As a result for the FP6 EUPASS project Lohse (2006) defined the connection between processes and resources for modular assembly systems. FP6 Pabadis'Promise (2006) project resulted in a manufacturing ontology (P2 ontology) and, reference architecture focusing on factory floor control.

Borgo & Leit (2007) developed the ADACOR ontology for distributed holon-based manufacturing focusing on processes and system interaction descriptions. ADACOR was later extended with an upper ontology Descriptive Ontology for Linguistic and Cognitive Engineering (DOLCE). Research done in the FP6 IP-PiSA project resulted an ontology, called Core Ontology, for connecting product, process and system domains under one reference model, Lanz (2010). The main goals these approaches generally try to achieve are: improved overall access to domain knowledge and additional information. However, none of these developed ontologies fully consider the needs above their narrow domain, Ray (2004), Lanz (2010), Jarvenpaa et al. (2010).

Ray (2004) introduced a roadmap from common models of data to self-integrating systems. The table 2 shows the 4 levels of representation. The table shows the logical steps for reaching first the creating of meaningful models (as in computational sense) to achieving finally systems that can autonomously exchange knowledge and operate based on shared knowledge.

According to the guidelines envisioned by Ray (2004), Lanz (2010) developed a common knowledge representation (KR) and semantics, called as Core Ontology, that allowed different design tools to interoperate across the design domains. The structure of the KR was formed on the basis of the requirements set by the knowledge management and integration challenges between different design tools, and the requirements set by the dynamic and open production environment. The developed model formalized the knowledge representation between product, process, and system domains utilizing fractal systems theory as a guideline. The surrounding system, be it the design environment or adaptive production system, can focus on the reasoning at different levels of abstraction, while the KR remained neutral for these reasoning procedures, but force the saved information to be consistent across the models.

Table 2: The evolution of representational power towards formal semantics, and the systems integration capabilities that could follow (Ray, 2004)

LEVELS OF REPRESENTATION	DESCRIPTION OF CHARACTERISTICS
Common Models of Data	In the lowest level the current state of the art, where the XML-based standards are utilized with relative ease within the IT sector, but not fully utilized in more conservative industry sectors.
Explicit and Formal Semantics	The second step, formal semantics, offers the generation of standardized representation that is formal enough to be parsed with computers.
Self-describing Systems	The third step is self-describing systems, where the systems can provide formal descriptions of their content and interfaces. This requires a formal semantic definition language that is rigorous enough to support logical inference.
Self-integrating Systems	The fourth level that Ray (2004) proposes is self-integrating systems. These systems are intelligent enough to be able to ask others for a description of their interfaces and, on the basis of the information thus acquired, adjust their own interfaces to be able to exchange information.

This approach differs from the traditional approaches by the fact that these tools are all utilizing already existing information as well as contributing specific information to the same model from different perspectives. The main objective of the developed KR was to achieve level two of the knowledge roadmap illustrated in the table 2. Other similar models exist, which utilize the complex nature of production systems. Most of these approaches are in the field of autonomous systems and control science.

TOWARDS MODULAR KNOWLEDGE ARCHITECTURE FOR THE DYNAMIC ENVIRONMENT

Understanding the Life-cycles of systems

All systems have their own life-cycles. In an open and complex operation environment the life-cycles play a very important role. The life-cycles that products can have are the most well-known life-cycle phases. These are such as in design, approved, in manufacturing, obsolete and such. These life-cycle phases represent the status of the design information. The resource units also have their own specific life-cycle phases. These life-cycle phases describe the essential part of individual system units. Fore example, in the case of manufacturing resources the payload of a robot, accuracy of the tool and joints or tolerances do change over the life-cycle of the machine. It may happen that the capabilities of a system decline when it proceeds along its life-cycle. An example could be the capability for manufacturing certain surface with tight tolerances is possible when the machine is relatively new, but once the operating hours exceed a certain level the capability to reach the needed tolerances is no longer possible. Another example is the combined capability of an advanced manufacturing center and its operator. The machine may have dormant capabilities to perform advanced operations, which can be obtained once the operator has achieved needed knowledge in this particular case. Now the combined system's capabilities have increased.

In the case of modular knowledge architecture, ICT has also its own life-cycle. It is accepted from the start that the business field may change. When the change happens the ICT architecture must also adapt to the change. The change can happen also in the technological side when new technologies replace old ones. This means that some of the services may become obsolete and new services need to be added. In order to keep the architecture maintainable one solution is to offer independent service modules that operate over one information model without direct integration to the underlying databases.

Layers of operation

One of the approaches divides the knowledge management system into three separate layers: databases, semantic operation logic (the knowledge representation) and services that utilize commonly available knowledge. The modular approach in ICT allows also the software vendors to enhance their production to be more modular and configurable thus allowing the service oriented operation model to be realized. Once the storing method is extracted from the logic and services, the new concepts can emerge. It is also seen that vendors can make new business strategies based on new modular system solutions and configuration of those instead of highly tailored solutions, which cannot be re-used later on.

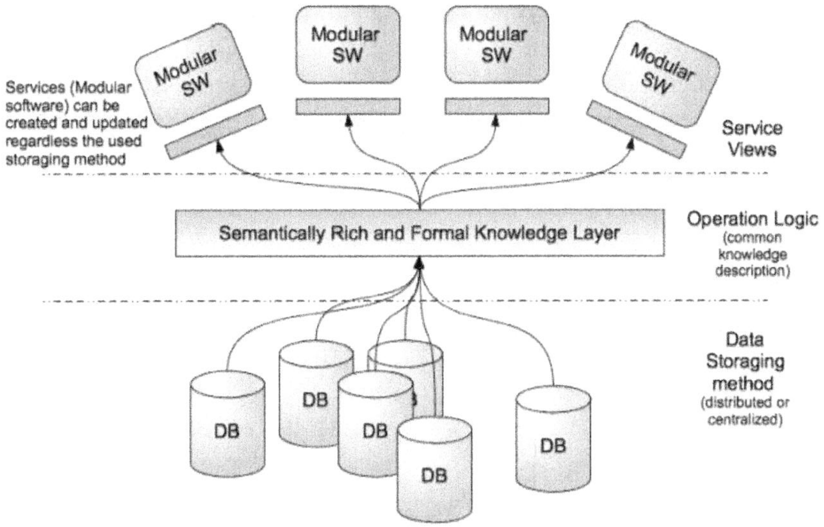

Figure 2: Modular ICT

The ultimate goals in this particular research effort were to provide an information architecture, which allows different utilization of domain knowledge, while keeping the core information consistent and valid throughout the life-cycles of that particular set of information. The primary requirements that were defined together with industry are:

- The model needs to represent the function of products and systems;
- The model needs to connect different domains under one representation;
- It must contain the history of changes applied to different instances;
- The model must serve as an input source for automated information retrieval and reasoning in the traditional and in holon-based operation environment;
- The model must be independent of the database implementation and services;
- The model must allow as well as facilitate the generation of different services; and
- The model must be extendable without disrupting the validity and consistency of the core domains.

IMPLEMENTATION OF A MODULAR ICT SYSTEM

The developed system, used here as an example, was based on the common knowledge representation and modular services would look as illustrated in figure 3. The clients contributing to the knowledge base are both commercial and university built existing systems and beta versions. Each of these tools requires specific domain related information and by processing the information they provide a set of services. However, the core of the system, the Knowledge Base (KB), needs to be extended to allow the capture and storing of semantically richer knowledge Lanz (2010) and Jarvenpaa et al. (2011).

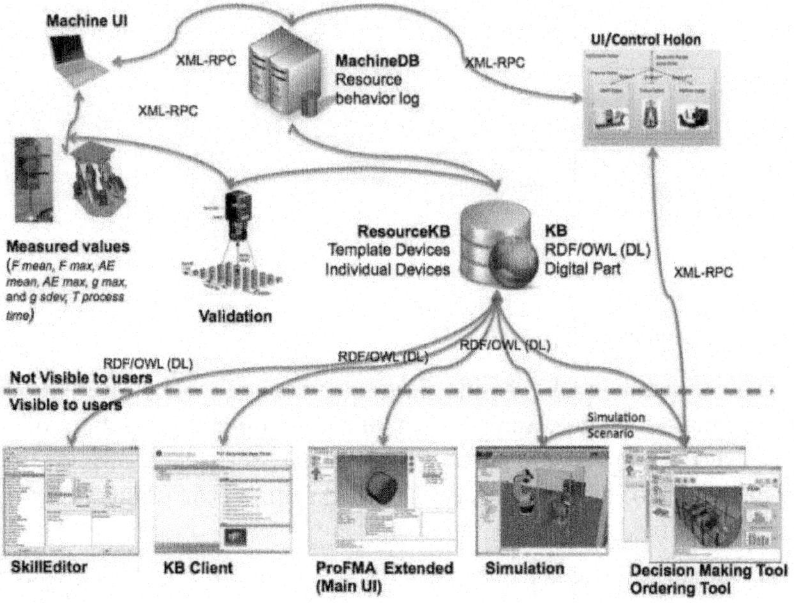

Figure 3: implementation

The utilized knowledge representation (KR) can capture the meaning of classes via relationships that are defined between the classes. This technology allows semantic richness to be embedded into the model. Several service providers can use the meaning of stored information for their own specialized purposes. The model is divided into three separate layers as illustrated in figure 2. By dividing the data reserves, operation logic and services into separate layers connected with interfaces the upgrading of layers becomes independent of each others. This allows services to be extended, replaced and modified throughout their life-cycles.

In this case study the whole system architecture, illustrated in figure 3 has several different interoperating software modules each providing one or two essential functions for the whole holonic manufacturing system. The architecture is designed in such way that each of the modules can be replaced with a new module if needed. The connection of the modules is mainly based on the shared information model, the Core Ontology, described in detail in Lanz (2010), Lanz et al. (2011) and in Jarvenpaa et al. (2011).

The tools in the environment are designed by keeping the modularization principles in mind. Each of the tools are contributing their specific information to the common information model. The tools provide one or two main functionalities to the software environment. The modular design of the software allows changes to be applied to the tools with minimum disturbances. For example the holon user interface (UI), which controls the actual production can be replaced with a commercial tool that provides queueing functionality for the system.

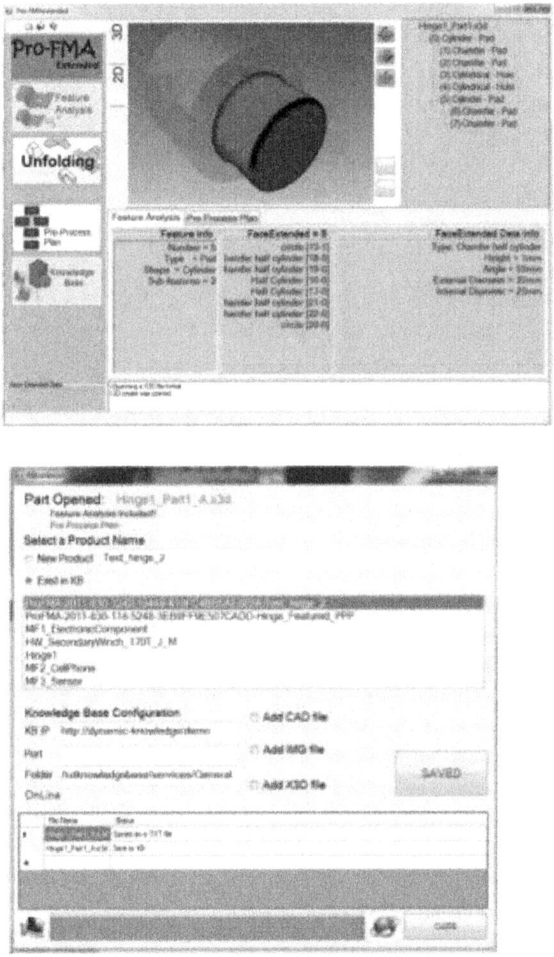

Figure 4: Pro-FMA tool

The tools are:

Content creation: Pro-FMA illustrated in figure 4 is used to define the product requirements from the product model given in virtual reality modeling language (VRML) or eXtensive 3D (X3D) format. Product requirements are those product characteristics or features that require a set of processes for product to be assembled or manufactured. Features can be geometrical or non-geometrical by nature. These processes are executed by devices and combination of devices possessing adequate functional capabilities, Garcia et al. (2011).

Context creation: The Capability Editor, illustrated in figure 5, allows user to add devices to the ontology and assign them capabilities and capability parameters and enables creating associations between the capabilities. In other words it creates rules about which simple capabilities are needed to form combined capabilities, Jarvenpaa et al. (2011).

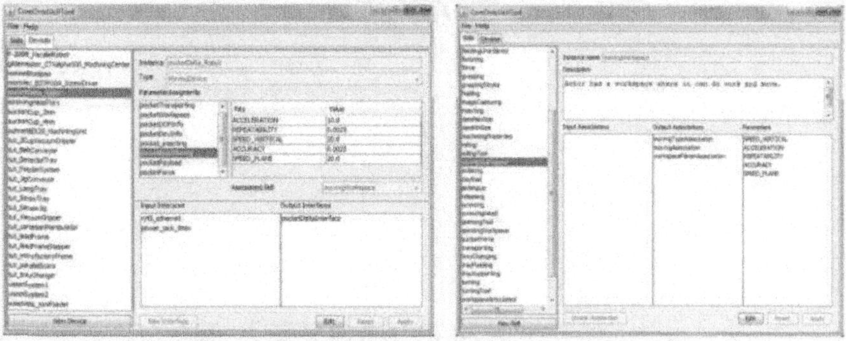

Figure 5: Editor for Capabilities

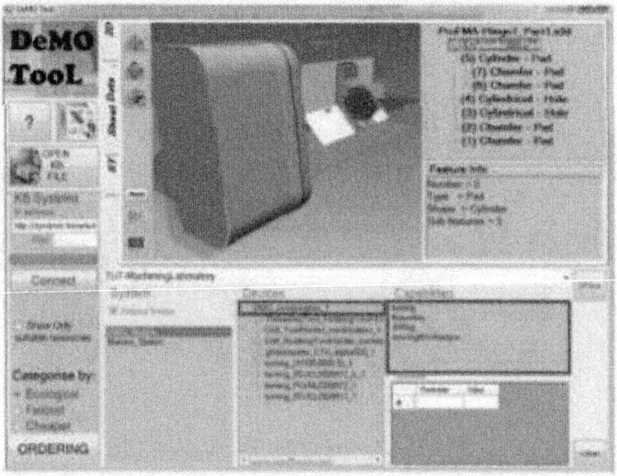

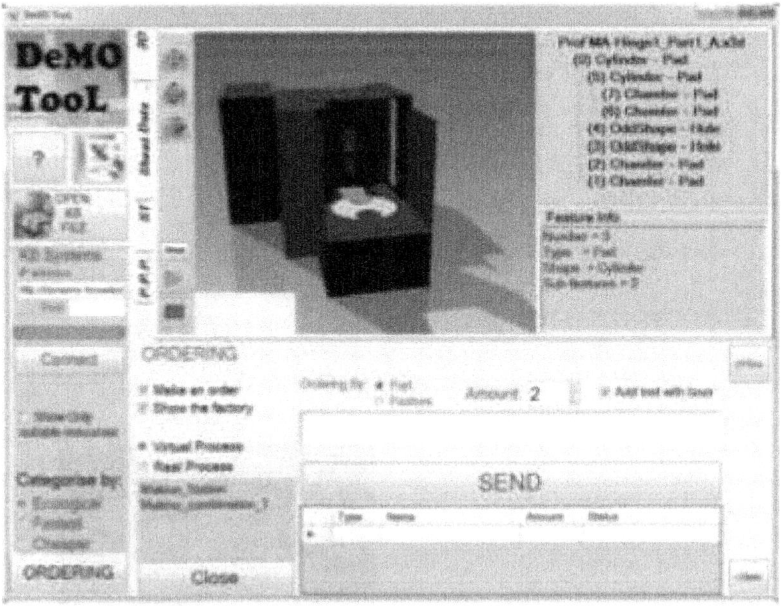

Figure 6: Decision Making and Ordering Tool

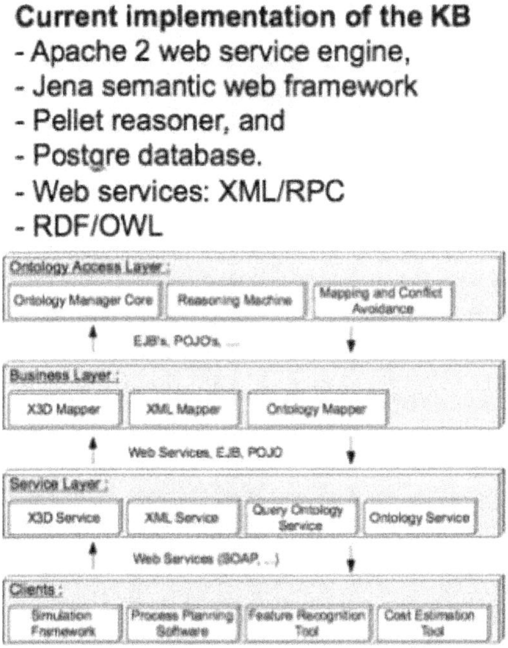

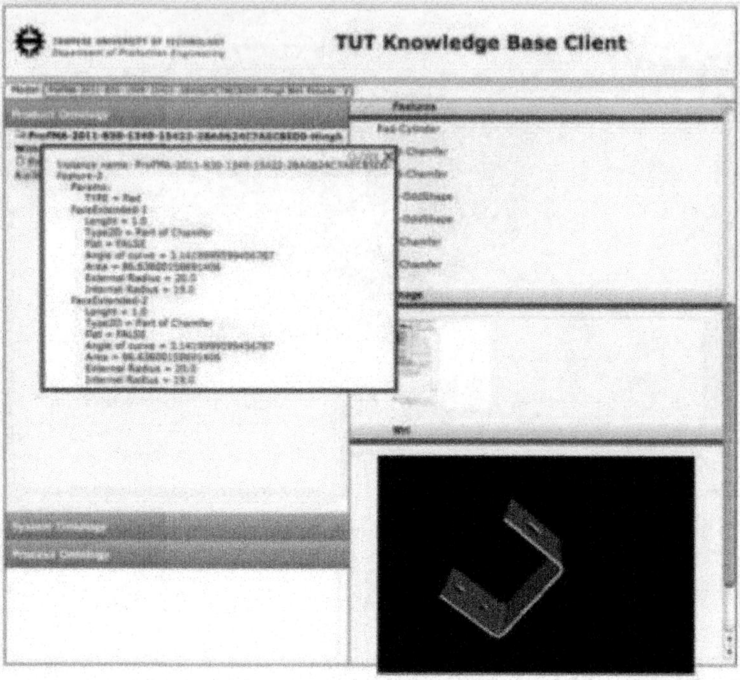

Figure 7: Knowledge Base and Knowledge Base Web Client

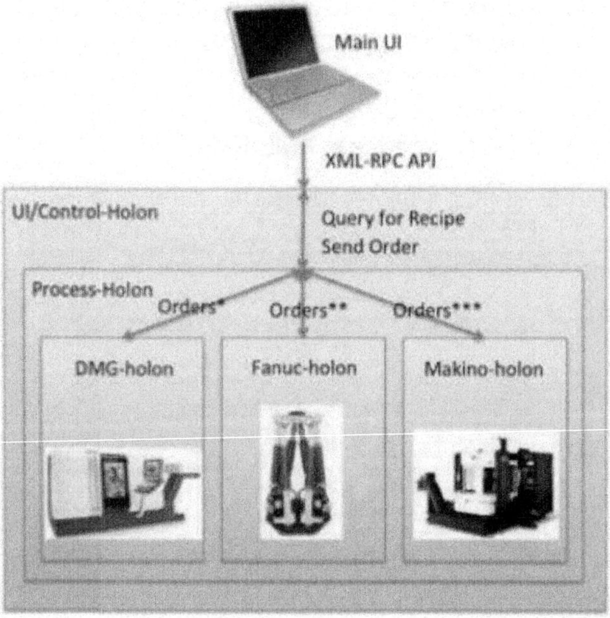

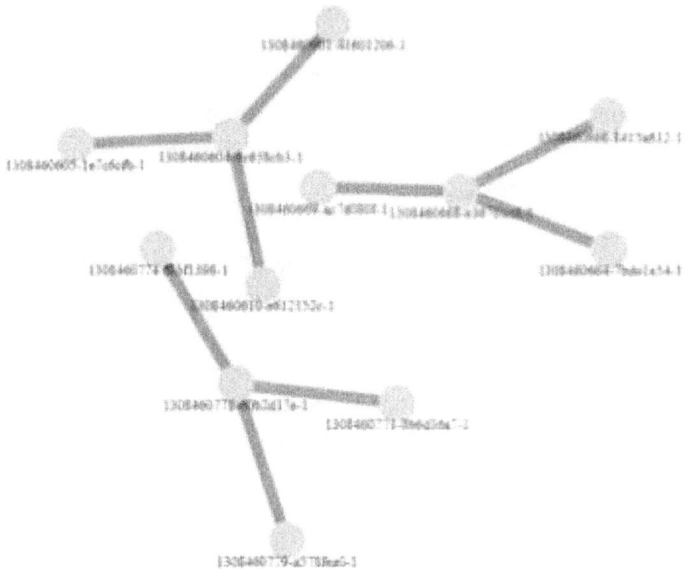

Figure 8: Holonic control based on Kademlia, the right side of the figure shows the messages sent between holons

Verification: A simulation tool is used for creating the manufacturing or assembly scenarios. Since the environment is holonic by nature, it is accepted that the simulation only expresses possible solutions. The operation principle inside the simulation also follows holonic guidelines. This means that part or product is routed to the first available and capable cell.

Ordering: The Decision Making and Ordering Tool (DeMO tool), in figure 6, is used for setting up orders in this environment. The tool supports the viewing of the simulation as its minor function. The main function of the DeMO tool is to verify the connection to the factory floor and forward the orders to the holonic UI, Garcia et al. (2011).

Common Knowledge Representation: The KB and ResourceKB, shown in figure 7, store the information created by Pro-FMA, Capability Editor and DeMO tool. This system serves also as reference architecture, since it can handle closed models as references. The knowledge representation used in this case is based on OWL DL. The simulation model can be attached to product definition if needed. Similarly closed sub-programs and Computer-Aided-x (CAx) models can be associated with the part/product/resource description, Lanz (2010).

Content Verification: A web-based KB client, shown in figure 7 is used for human friendly information browsing. This tool serves as product data management (PDM) system's web-based user interface (UI). The client allows only limited set of changes to be applied to the ontology. These changes are for example a new name for a product, part or other instance. For more details, please see Lanz (2010).

Operations Management: The process flow and distribution of tasks to each manufacturing or assembly cell is done with the Control Holon, see figure 8. The control holon observes the status of the system and available capabilities of system units (manufacturing resources in this case). The manufacturing resources can enter to and leave from the network without disturbing the whole system. This holonic control system distributes the tasks to suitable and available cells or stations based on the capability requirements defined by Pro-FMA earlier.

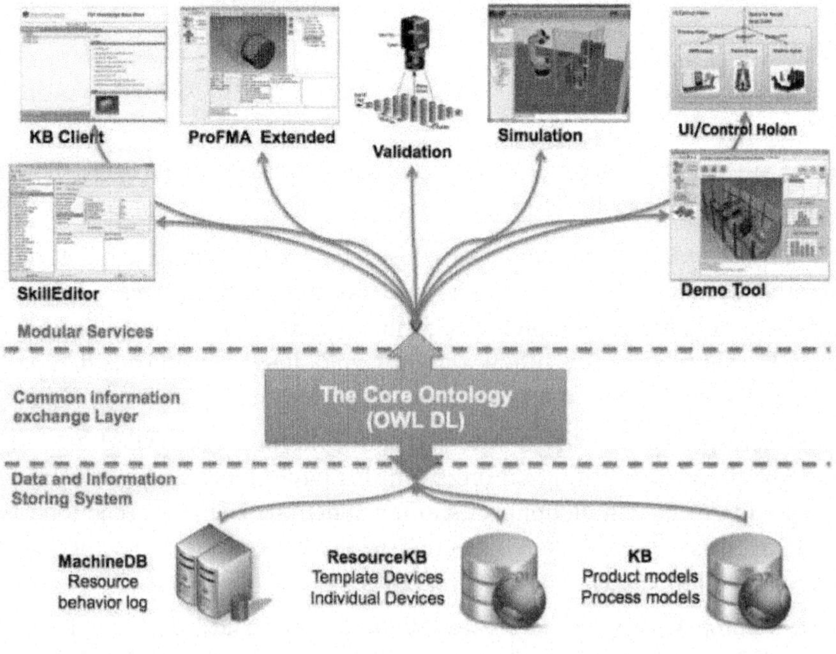

Figure 9: The implementation is formed based on the modular ICT concept

The tools are divided into the layers described in previous chapter, see figure 2. The implemented environment, in figure 9, allows the addition of new services which can contribute and /or utilize already existing information, thus proving the modular ICT concept feasible for an adaptive, open and complex manufacturing environment. These tools constitute the necessary core for a modular system. There are additional services that could be added to this environment. These are traditional operation management module for production orchestration and machine vision based validation module. Both services are seen as extra for the core system.

CONCLUSION

Manufacturing after all is the backbone of each and every society, and in order for a society to be sustainable in long run the manufacturing has to be sustainable as well. From another point of view, manufacturing systems are shifting from being to becoming. This means that as the intelligence and cooperativeness advances the system will become a society where the rules, possibilities and constrains of a society as we know it will also apply. In order to achieve goals in the manufacturing society this research effort will contribute tremendous assets for securing the paradigm shift while keeping the manufacturing industry sustainable, flexible and adaptive. Without acceptance, further concept developments and implementation of the open and complex system approach the industry will not meet the challenges of the evolving environment.

It is seen that one partial solution is to develop these kind of modular ICT architectures that support the evolution of systems. However, it is understood that there is a lot of developments and solutions needed, since the industry cannot adopt partial solutions. Industry will require a concept that allows several data sources to be combined under one coherent and valid representation that facilitate the design and utilization of intelligent services in open and dynamic operations environment.

This paper introduced the context and operation principles of a dynamic system, and what is needed to support this kind of system from the knowledge management perspective. The article emphasized the challenge of dynamic systems from the life-cycle perspective as well, since all of the system parts be those software or hardware have their specific life-cycle phase. The division of architecture does provide tremendous possibilities for service development in future. As a proof of concept one type of modular ICT architecture and its core tools were introduced.

These results introduced here can be utilized in other fields than manufacturing engineering as well. The field of constructed environment and urban development has already seen the potential of an open world system where the input can be delivered in formal representation and services can be created independently of each other.

DISCUSSION

When discussing about the holonic concepts with different people in seminars, workshops and conferences, a common comment/question has been: "Holonic manufacturing systems were developed 20-30 years ago and they didn't work then. How could they work now?" Shortly put, the answer could be technological and methodological development of knowledge and information management. Reasoning needed in the holonic systems relies on information and knowledge. Even though the concept of holonic manufacturing has remained similar throughout the years, information technology has made huge leaps enabling the implementation of these concepts in a feasible way. The novel methods to manage and distribute knowledge, such as semantic web and web service technologies, as well as semantic knowledge management systems, have been paving the way for the successful implementation of holonic systems.

Another question, which often arises in discussions has been: "Why holons? What advantages we gain by implementing holonic architecture? The implementation seems to be a huge task." Holons are autonomous and self-describing entities having well defined interfaces and the ability to communicate and co-operate with other holons. The modularity and self-organization ability enables the holonic systems to be extendable and adaptable. New holons, be they software system modules, new manufacturing resources or human workers, can enter and leave the system without disturbing the operation of the whole system. Each holon, module, knows its own purpose and the inputs and outputs, making the operation more transparent. In a holonic system it is possible to make changes in individual modules without the need to change/re-program the whole system. Until recently the holonic paradigm has only been implemented to physical devices and immediate control architecture of those. The design, operations management and supporting ICT systems have been ignored. However, as the ICT is expected to adapt to the changes in the production environment the holonic paradigm provides operation principles for this side as well.

Manufacturing is not the only domain, where the holonic paradigm could be applied. Actually, it could be applied almost anywhere, like in a medical and logistical domains. A good example can be found from city logistics. Cities, and their design, are not centrally controlled organized systems, but they are characterized by some level of chaos and the continuous threat of the chaos to expand to other operational areas. This chaos is controlled by hierarchical control systems where the control is coming from the top. From this viewpoint chaos is always considered as a negative element. This kind of systems need always be implemented as closed systems in order to prevent chaos.

The problem here is that innovations do not happen in order and harmony. The innovation always causes temporary chaos. Hierarchical control naturally strangles innovation. Therefore, what is needed is a control system where chaos is not a matter of crisis, but a normal event the system can handle in a flexible and efficient way. This kind of control system can be called as "chaordic system" (chaos + order). "Chaordic system" is self-organizing system which can always find a new equilibrium when the situation changes. The holonic control architecture can answer to the requirements of the "chaordic system". This idea has been presented to experts in the field of city logistics with very good feedback. The experts saw significant development potential for their business in holonic architecture in ICT and following the "open system" principles. However, all of this will be just theoretical discussion unless the surrounding ICT does support the change.

REFERENCES

1. Awad, E.M. & Ghaziri, H.M. (2004).Knowledge management, Upper Saddle River, NJ, Pearson Education Inc.
2. Bock, C. & Gruninger, M. (2005). PSL: A Semantic Domain for Flow Models, Journal of Software and Systems Modeling, 4:2
3. Borgo, S. & Leit, P. (2007). Foundations for a core ontology of manufacturing, Integrated Series in Information Systems, vol 14
4. Chavalarias, D.; Cardelli, L.; Kasti, J. et al., (2006).Complex Systems: Challenges and opportunities, an orientation paper for complex systems research in fp7, European Commission
5. Cotsaftis, M. (2009). A passage to complex systems, in Complex Systems and Self-organization Modeling, C. Bertelle, G. H. E. Duchamp, and H. Kadri-Dahmani, eds., Springer, 2009 European Commission (2003). Working Document For The MANUFUTURE 2003 Conference Gruver,
6. W. (2004). Technologies and Applications of Distributed Intelligent Systems, IEEE MTT-Chapter Presentation, Waterloo, Canada
7. Dept. of Homeland Security (DoHS), NAT, Cyber Security Division: Catalog of Control Systems Security, (2008). Recommendations for Standards Developers FP6 Pabadis'Promise 2006. D3.1 Development of manufacturing ontology, project deliverable, The PABADIS'PROMISE consortium
8. Garcia, F.; Jarvenpaa, E.; Lanz, M. & Tuokko, R. (2011). Process Planning Based on Feature Recognition Method, Proceedings of IEEE International Symposium on Assembly and Manufacturing (ISAM 2011), 25-27th of May, 2011, Tampere, Finland

9. Iria, J., (2009), Automating Knowledge Capture in the Aerospace Domain, Proceedings of K-CAPÕ09,Redondo Beach, California, USA
10. Jarvenpaa, E.; Lanz, M.; Mela, J. & Tuokko, R. (2010). Studying the Information Sources and Flows in a Company Ð Support for the Development of New Intelligent Systems. Proceedings of the FAIM2010 Conference, July 14-17, 2010 California, USA, 8 p.
11. Jarvenpaa, E.; Luostarinen, P.; Lanz, M.; Garcia, F. & Tuokko, R. (2011).Presenting capabilities of resources and resource combinations to support production system adaptation, Proceedings of IEEE International Symposium on Assembly and Manufacturing (ISAM 2011), 25-27th of May, 2011, Tampere, Finland
12. Jarvenpaa, E.; Luostarinen, P.; Lanz, M.; Garcia, F. & Tuokko, R. (2011). Dynamic Operation Environment Ð Towards Intelligent Adaptive Production Systems, Proceedings of IEEE International Symposium on Assembly and Manufacturing (ISAM 2011), 25-27th of May, 2011, Tampere, Finland Koestler, A. (1968).
13. Ghost in the Machine, Penguin, ISBN-13: 978-0140191929, p. 400
14. Krima, S.; Barbau, R.; Fiorentini, X.; Sudarsan, R. & Sriram, R.D., 2009, ontoSTEP: OWL-DL Ontology for STEP, NIST Internal Report, NISTIR 7561
15. Lanz, M.; Lanz, O.; Jarvenpaa, E. & Tuokko, R. (2010). D1.1 Standards Landscape, KIPPcolla: internal project report Lanz, M. 2010. Logical and Semantic Foundations of Knowledge Representation for Assembly and Manufacturing Processes, PhD thesis, Tampere University of Technology
16. Lanz,M.; Rodriguez, R. & Tuokko,R. (2010) Neutral Interface for Assembly and Manufacturing Related Knowledge Exchange in Heterogeneous Design Environment, published as book, Svetan M. Ratchev (Ed.): Precision Assembly Technologies and Systems, 5th IFIP WG 5.5 International Precision Assembly Seminar, IPAS 2010, Chamonix, France, February 14-17, 2010. Proceedings. IFIP 315 Springer 2010, ISBN 978-3-642-11597-4, France
17. Lanz, M.; Jarvenpaa, E.; Luostarinen, P.; Tenhunen, A.; Tuokko, R. & Rodriguez, R. (2011). Formalising Connections between Products, Processes and Resources Ð Towards Semantic Knowledge Management System, Proceedings of Swedish Production Symposium 2011, Sweden
18. Lohse, N. (2006), Towards an Ontology Framework for the Integrated Design of Modular Assembly Systems, PhD thesis, University of Nottingham ROSETTA Project Presentation May 2010. FP7 ROSETTA

Project consortium, URL: http://www.fp7rosetta.org/public/ ROSETTA_ Project_ Presentation_May_2010 _webpage.pdf

19. Ray, S. (2004). Tackling the semantic interoperability of modern manufacturing systems, in Proceedings of the Second Semantic Technologies for eGov Conference Salminen, K.; Nylund, H. & Andersson,P. (2009). Role based Self-adaptation of a robot DiMS based on system intelligence approach, proceedings of Flexible Automation and Intelligent Manufacturing (FAIM 2009), US
20. Tharumarajah, A.; Wells, A.J. & Nemes, L. (1996). A Comparison of the Bionic, Fractal and Holonic Manufacturing Concepts, International Journal of Computer Integrated Manufacturing, vol.9, no.3/1996
21. Ueda, K.; Vaario, J. & Ohkura, K. (1997). Modelling of Biological Manufacturing Systems for Dynamic Reconfiguration, Annals of the CIRP, 46/1: 343-346/1997
22. Valckenaers, P.; van Brussel, H.; Bongaerts, L. & Wyns, J. (1994). Holonic manufacturing execution systems, CIRP Annals - Manufacturing Technology, nro 54/1994

CITATION

CHAPTER 1

Ernesto López-Mellado, Agent-Based Synthesis of Distributed Controllers for Discrete Manufacturing Systems, doi: 10.4236/jsea.2011.43015.

CHAPTER 2

Yihai He, Zhenzhen He, Linbo Wang, and Changchao Gu, "Reliability Modeling and Optimization Strategy for Manufacturing System Based on RQR Chain," Mathematical Problems in Engineering, vol. 2015, Article ID 379098, 13 pages, 2015. doi:10.1155/2015/379098

CHAPTER 3

Jilcha K, Berhan E, Sherif H (2015) Workers and Machine Performance Modeling in Manufacturing System Using Arena Simulation. J Comput Sci Syst Biol 8:185-190. doi: 10.4172/jcsb.1000187

CHAPTER 4

Vogel-Heuser, B. (2014) Usability Experiments to Evaluate UML/SysML-Based Model Driven Software Engineering Notations for Logic Control in Manufacturing Automation. *Journal of Software Engineering and Applications*, 7, 943-973. doi: 10.4236/jsea.2014.711084.

CHAPTER 5

Singh, R, Singh, R. and Khan, B. (2015) A Critical Review of Machine Loading Problem in Flexible Manufacturing System. *World Journal of Engineering and Technology*, **3**, 271-290. doi: 10.4236/wjet.2015.34028.

CHAPTER 6

Y. Lee, Y. Jhan, C. Chung and Y. Hsu, "A Prediction Method for In-Plane Permeability and Manufacturing Applications in the VARTM Process," *Engineering*, Vol. 3 No. 7, 2011, pp. 691-699. doi: 10.4236/eng.2011.37082.

CHAPTER 7

J M Edwards and I A Coutts, The flexible integration of machine objects within distributed manufacturing systems, http://iopscience.iop.org/article/10.1088/0967-1846/2/4/001/pdf

CHAPTER 8

Faieza Abdul Aziz, Izham Hazizi Ahmad, Norzima Zulkifli and Rosnah Mohd. Yusuff (2012). Particle Reduction at Metal Deposition Process in Wafer Fabrication, Manufacturing System, Dr. Faieza Abdul Aziz (Ed.), ISBN: 978-953-51-0530-5.

CHAPTER 9

C. Wheeley, P. Mago and R. Luck, "A Comparative Study of the Economic Feasibility of Employing CHP Systems in Different Industrial Manufacturing Applications," *Energy and Power Engineering*, Vol. 3 No. 5, 2011, pp. 630-640. doi: 10.4236/epe.2011.35079.

CHAPTER 10

Jorge Cortes, Ignacio Varela-Jimenez and Miguel Bueno-Vives (2012). Reconfigurable Tooling by Using a Reconfigurable Material, Manufacturing System, Dr. Faieza Abdul Aziz (Ed.), ISBN: 978-953-51-0530-5, InTech, DOI: 10.5772/32560.

CHAPTER 11

Minna Lanz, Eeva Jarvenpaa, Fernando Garcia, Pasi Luostarinen and Reijo Tuokko (2012). Towards Adaptive Manufacturing Systems - Knowledge and Knowledge Management Systems, Manufacturing System, Dr. Faieza Abdul Aziz (Ed.), ISBN: 978-953-51-0530-5.

INDEX